W9-CWV-971

PRIMER OF NEUROLOGY AND NEUROPHYSIOLOGY

James P. Chaplin, Ph.D.

Department of Psychology
St. Michael's College
Winooski, Vermont

Aline Demers, R.N., M.S.

Department of Professional Nursing
University of Vermont
Burlington, Vermont

A Wiley Medical Publication

JOHN WILEY & SONS
New York • Chichester • Brisbane • Toronto

Library of Congress Cataloging in Publication Data:

Chaplin, James Patrick, 1919-
Primer of neurology and neurophysiology.

(A Wiley medical publication)
Bibliography: p. 249
Includes index.
1. Neurology. 2. Neurophysiology. I. Demers, Aline, joint author. II. Title. [DNLM: 1. Nervous system—Anatomy and histology. 2. Nervous system—Physiology. WL101.3 C464p]
RC346.C45 616.8'04'7 78–6680
ISBN 0-471-03027-9

Printed in the United States of America

10 9 8 7 6 5 4 3 2 1

PREFACE

This primer of neurology and neurophysiology has been written for undergraduates enrolled in courses that emphasize the structure and functions of the nervous system. Because of its clinical applications, it should prove particularly useful for students in nursing and allied health sciences. Graduate nurses and others in the medical field who wish to review basic neurology and neurophysiology will also find it helpful.

In preparing the text, every effort was made to clarify and simplify, without sacrificing accuracy, the intrinsically difficult anatomical relationships and physiological processes that make up neuroanatomy and neurophysiology. In pursuit of this objective, we begin with the simplest unit of the nervous system, the neuron, and progress gradually through the spinal pathways to structures of the hindbrain, midbrain, and forebrain. Finally, we consider the anatomy and functions of the cerebral cortex. Wherever possible, we have attempted to make each chapter independent, not hesitating to review or summarize previously considered material on the assumption that students meeting the nervous system for the first time need repetitive reinforcement in order to master the many technical terms and anatomical relationships involved.

We have also made an effort to write from a humanistic point of view as reflected in the style of writing and in the description of classic experiments in neurophysiology and illustrative case histories. Toward this same end we have included the Greek and Latin derivations of anatomical terms as these occur in the text in order to help the student better understand their often obscure connotations.

Careful attention and emphasis has been given to the diagrams and illustrations in this text. These were developed especially for this volume by one of the authors (A.D.) from basic anatomical sources in order to ensure the closest possible correlation between textual material and illustrations. A special effort has been made to clarify those areas that our experience has revealed to be particularly difficult, such as the spinal pathways and the autonomic nervous system.

As is customary in primers, we have made no attempt to cite references or bibliographic material in the text. However, for those who may wish to explore the nervous system at a more advanced level, an annotated bibliography of suggested readings has been appended.

Finally, we would like to take this opportunity to express our appreciation to those who helped in the preparation of this book. Rodney Parsons of the Department of Physiology of the University of Vermont provided a detailed criticism of the entire rough draft and made numerous valuable suggestions. The manuscript, in whole or in part, was also reviewed by Marjorie A. Miller and Vera Stolar of Cornell University–New York Hospital School of Nursing. We are grateful for their suggestions. Barry Krikstone of the Department of Psychology, St. Michael's College, provided numerous source materials.

Special acknowledgment is made to the Medical Center Hospital of Burlington, Vermont, for providing case history material and to Mary Hill, Director of Medical Records, who so skillfully retrieved it from thousands of medical records.

J. P. C.
A. D.

CONTENTS

1
INTRODUCTION TO THE NERVOUS SYSTEM

A SPACE ODYSSEY

In an age when awe-inspiring feats of science and technology are commonplace, when planets hundreds of millions of miles distant lie within man's reach, and when the history of the birth and death of suns millions of light years from earth is being written, it is tempting to think that all the mystery and excitement of great discovery has at last yielded to man's insatiable curiosity. However, there remains what may be the greatest mystery of all—the human nervous system. Every research scientist and clinician who has studied the brain of man knows the challenge and the adventure of discovery that remain hidden within the complexities of those few thousand grams of tissue that Sir Charles Sherrington, the eminent British neurologist, called "the great ravelled knot."

During the long history of evolution, living tissues differentiated into many forms, some simple, some incredibly complex in design, but none can rival the nervous system in intricacy of form and function. The surgeon can transplant the relatively uncomplicated mass of muscle that we call the heart. Or a patient can be connected to a respirator or kidney machine and his digestive system can be by-passed by intravenous feedings, but no one has yet discovered a substitute for the brain. Indeed, as clinicians push the frontiers of death further and further back with the help of such sophisticated equipment, there is one frontier that marks the final boundary—an unresponsive brain.

The cerebral cortex alone—a quarter-inch thick layer of gray matter that is laid over the cerebrum like a wrinkled mantle—contains approximately 9,000,000,000 cells. A famous neurologist of the past generation, C. Judson Herrick, calculated that if only one million cortical cells interacted with each other in pairs in every possible way, the total number of potential interactions would have to be expressed by the figure $10^{2,782,000}$. This number can only be compared to those that are used in describing the very fabric of the universe, and even here the analogy falls short. It is estimated that there are 10^{56} *atoms* in our solar system and perhaps 10^{66} in the universe. But even as we ponder the awesome magnitude of these numbers, we must remember that it is not a million cells that are capable of interacting with each other in the cerebral cortex but hundreds of millions. Moreover, each of those hundreds of millions of cells of neural tissue is potentially capable of interacting with every other in many more possible ways than one or two.

And finally, another way of appreciating something of the marvel of the human brain is to ask what kind of computer could rival its complexity. The absolute answer is, of course, none, but a fair beginning might be made by a machine the size of the Empire State Building filled with electronic tubes and transistors that required a power supply equal to the entire hydroelectric output of Niagara Falls.

IN THE BEGINNING . . .

Evolutionarily speaking, the nervous system originated in the single-celled animals, since they show irritability and conductivity, those two basic functions of all nervous tissue. If the amoeba (a single-celled protozoan whose kith and kin have given much to biological research) is lightly stimulated with an irritant anywhere on its cell membrane, the entire organism slowly flows away from the source of stimulation. Similarly, another single-celled protozoan, the paramecium, which is better equipped for locomotion than the amoeba, will quickly dart away from an irritating stimulus anywhere on its membrane. In these forms the entire organism shows irritability and conductivity. There is no specialization of function.

With the development of multicellular organisms came specialization of function, and those cells that specialized in irritability evolved into receptors or sense organs, while those whose specialty was conductivity became neurons (*nevus,* nerve, sinew) or fibers that make up nerves. Some of these neurons specialized in conducting impulses to the central nervous system and so are called sensory neurons. Other neurons specialized in conduction away from the nervous system to muscles and glands and are called motor neurons. As organisms evolved into highly complex forms, association neurons were needed to interconnect sensory and motor fibers and to serve as the mediators of reflex action and habit, and, in the higher forms, the cognitive processes involved in perceiving, thinking, and reasoning.

OUNCES TO POUNDS: THE BRAIN FROM FISH TO MAN

If we take a quick journey up the phylogenetic scale from primitive fishes to man, pausing here and there to compare the brains of selected species, we will become familiar with some of the major parts of the central nervous system, which consists of the brain and the spinal cord. We shall see that evolution did much to make the brain increasingly complex and specialized in the more advanced species but, at the same time, retained all of the older parts. This lack of radical change should not surprise us, for the environment that supports the phylogenetic kingdom is the same for all. What makes the nervous system of one species differ from that of another is the particular niche that species occupies in the world's ecology and the size and complexity of the body that the nervous system must control.

And so we find that the shark, a primitive fish whose chief receptors are olfactory in nature has an enormous proportion of its brain devoted to this sensory process (Fig. 1-1). Living as it does in a kind of diluted soup rich in odors, its chief activities center around catching prey for its famous jaws, and so it does not require cerebral hemispheres but only a well-developed midbrain to conduct impulses between its olfactory apparatus and motor centers. Another prominent and important area in the midbrain of sharks and all higher forms is the collicular (*collis,* mound) region that contains the inferior and superior colliculi, centers that mediate auditory and visual reflexes for these swift creatures of the sea.

In moving up from the fishes to reptiles—and it is a long step—we find that several noteworthy changes have occurred in the evolution of the brain. The alligator, who is our representative of the species, has a less prominently developed olfactory brain than the shark, presumably because fewer of its activities are guided by olfaction and more by its other senses. The cerebral hemispheres are present in reptiles, but are only poorly developed and comparatively small. There is no true cerebral cortex. Noteworthy among reptiles were the dinosaurs, whose gigan-

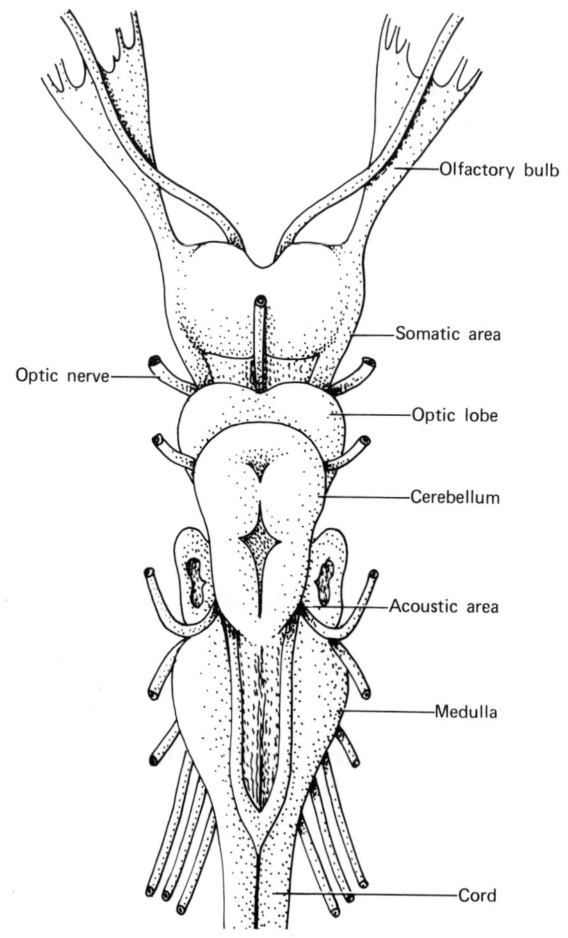

FIGURE 1-1. The brain of the shark. Note the relatively large size of the olfactory bulbs.

tic size required a sort of auxiliary brain located over the rear legs to aid in coordinating those huge appendages. One is tempted to draw an analogy between the dinosaurs and the hook-and-ladder fire truck, which needs a real backseat driver for the rear wheels as well as one for those up front.

In birds the cerebral cortex is present, and the cerebral hemispheres are much more prominent than in reptiles. The optic lobes are well developed, and the cerebellum, whose function it is to coordinate and control motor activities, is quite prominent—not surprising in view of the complex nature of locomotor activity in birds. The large optic centers we might also expect in this species, since many of their kind hunt small insects on the wing or, like the eagle, dive on their prey from great heights.

Before leaving the birds, we should note that this species is often maligned in the derogatory phrase, "bird brain." Let it be noted that, for its body weight, the robin has a comparatively large brain. Though the robin's brain weighs only a few grams whereas the whale's weighs fourteen pounds, the robin's is larger than the whale's in relation to body weight.

The dog, affectionately known as man's best friend, resembles him in having a true cerebral cortex with some wrinkling, a sure sign of increasing complexity in brains. Note that at this comparatively high level of evolution, the cerebrum is now the dominant part of the brain overshadowing all the rest. The older parts of the brain, such as the olfactory and optic lobes, are almost entirely hidden by the enormous development of the newer parts of the brain with its predominantly cortical tissue. This same trend is brought to its highest level of development in the primates in which the cerebrum, and particularly the cerebral cortex, so overshadows the rest of the brain that it is the only visible structure when the brain is viewed from the top. However, in looking at a generalized mammalian brain (Fig. 1-2), we see that all of the old structures are present but have been enveloped by the mushrooming cortex and its associated fibers of the newer portions of the brain. It is interesting to speculate on whether the evolutionary tendency toward cerebral and particularly frontal lobe dominance will continue—provided, of

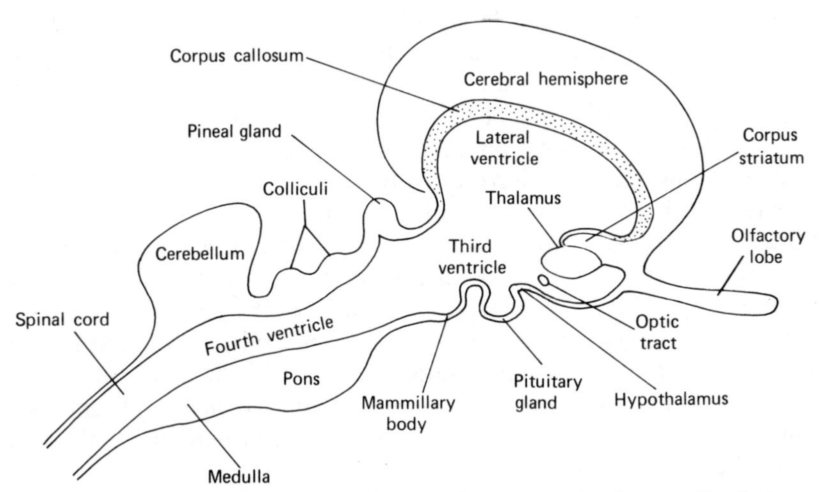

FIGURE 1-2. A schematic diagram representing the generalized mammalian brain.

course, man does not use his cerebrum to destroy the world he lives in and his own species with it. But assuming evolution is allowed to continue, men and women of the future may have gigantic heads with huge eyes and relatively tiny noses and ears, reflecting the comparative importance of the sense organs to future man and the continuation of the trend toward cerebral dominance.

THEY ALSO SERVE: NOTES ON THE EVOLUTION OF THE LOWER CENTERS

The spinal cord's structure and functions remain relatively fixed throughout the phylogenetic series. They are: conduction of nervous impulses from the bodily receptors to the brain and from motor centers in the brain to nerves originating in the cord that control muscles. The cord also serves as an integrating center for the control of reflexes. However, we should point to two important changes. First, those nerves and their associated centers that control the functions of the visceral organs (the autonomic functions; see Chap. 5), become separated off from the main sensory and from the motor tracts of the cord that control voluntary activities. Second, connections between the spinal cord and brain are shunted through the medulla in lower forms but in the more advanced species go directly to the thalamus (*thalos,* chamber), a collection of centers that lie deep within the forward portion of the brain and serve as important relay centers.

Similarly, the medulla oblongata (*medulla,* marrow; *oblongata,* oblong) is uniform in structure and function throughout the vertebrates. It contains centers for the regulation of respiration and heart rate and for the control of balance and spatial orientation. How much of the medulla is devoted to these functions varies with the species.

The midbrain, as we have already noted, contains the centers for reflexes associated with vision and audition; and in fishes, centers are provided for the lateral line organs. Lateral line organs are a series of cutaneous receptors found along the sides of some fishes and water-dwelling amphibians. They respond to currents of water and changes in pressure. These organs and their associated neural centers disappear in higher forms, and the relative importance of the colliculi as centers for vision and audition diminishes as we move from the birds and reptiles to man, in whom visual and auditory functions are largely under the control of the cerebral cortex.

The thalamus makes up most of the forebrain in the lower vertebrates. In man it is deeply buried under the cerebral cortex. In the lower forms the thalamus is primarily concerned with vision, but is more differentiated in higher forms and includes centers for hearing and the skin senses. However, as the cerebral cortex becomes dominant in the higher forms, particularly the primates, the thalamus is relegated to the role of a relay station between the cortex and lower centers.

The corpus striatum (*corpus*, body; *striatum*, striped), or striated body, is a collection of motor centers first observed in the fishes. It is found throughout the vertebrate series but is of increasingly less importance in the higher forms.

BACK TO THE BEGINNING . . .

As individuals we begin life as a single cell, the ovum, which is discharged from the cornucopia of the mother's ovary where all female reproductive cells are stored. Fertilized by the sperm or male sex cell somewhere in one of the Fallopian tubes, the egg, pushed along by ciliary action, begins a hazardous journey to the uterus dividing even as it goes. It implants in the uterus by attaching to the wall of that nutritive organ. Quickly, the future citizen divides into hundreds and thousands, then millions, of cells. As he or she—still an unrecognizable mass of cells—develops, growth takes place at different rates in various parts of the tissue mass. By looking through his microscope at early embryos prepared as slides, the embryologist can soon detect sheets or layers of cells that lie on the interior and exterior of the mass called the ectoderm (*ectos*, outside; *derma*, skin) or outside layer and the entoderm or inside layer. Shortly thereafter a third or middle layer, called the mesoderm, begins to differentiate. The ectodermal cells on the back or dorsal side of the embryo develop into the skin and nervous system. The mesodermal cells become the skeleton, muscles, connective tissues, and heart and blood vessels. The entoderm develops into the digestive and respiratory systems and associated glandular structures.

A HOLE IN THE HEAD: THE NEURAL TUBE

Early in the development of the mass of cells that make up the back of the embryonic layer, the tissues show a slight depression, the neural groove. Because the cells at the edge of the groove proliferate more rapidly than those in the center, the groove deepens and is forced inward eventually to become a separate tube. This process is shown diagrammatically in Figure 1-3. The entire nervous system develops from this primitive tube; and, despite all of the complex changes it undergoes, the adult nervous system retains its closed, tubular form. Our central nervous system is, then, hollow from head to tail, although filled with a teacup full of clear liquid known as the cerebrospinal fluid.

A moment's reflection will show that the head of the neural tube must undergo the most complex of the developmental changes that take place during embryonic and fetal life. And this, in fact, is precisely what occurs. Beginning about three weeks after conception, the head end of

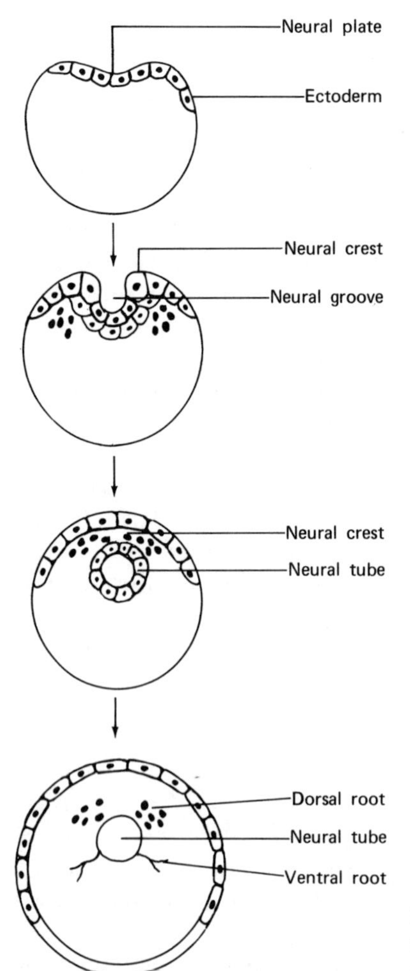

FIGURE 1-3. The formation of the neural tube and the ventral and dorsal spinal roots from the ectoderm.

the neural tube, which is destined to become the brain, shows three swellings or enlargements. These are the forebrain or prosencephalon (*pro,* forward; *enkephalos,* brain), the midbrain or mesencephalon (*mes,* middle), and the hindbrain or rhombencephalon (*rhomb,* a parallelogram). Figure 1-4 shows these vesicles or swellings diagrammatically.

Further changes soon result in additional subdivisions, and these, too, have been given names descriptive of their positions or appearance. The prosencephalon divides into the telencephalon (*tele,* far or distant) and diencephalon (*di,* divided). The mesencephalon remains undivided but the rhombencephalon becomes the metencephalon (*met,* after, back) and myelencephalon (*myelos,* marrow). The remainder of the neural tube develops into the spinal cord. Figure 1-5 shows the divisions of the embryonic brain and the more familiar parts that derive from them.

From the telencephalon the olfactory bulbs, cerebral hemispheres and corpus striatum develop. The diencephalon gives rise to the thalamus and hypothalamus (*hypo,* below), a center for the regulation of emotional and motivational behavior. The mesencephalon develops into the region of the brain containing the colliculi already described as reflex centers for vision and hearing. The metencephalon gives rise to the cerebellum (little brain) or that part of the brain that is concerned with the coordination and regulation of motor behavior, and its associated structure the pons (*pons,* bridge), which connects the two hemispheres of

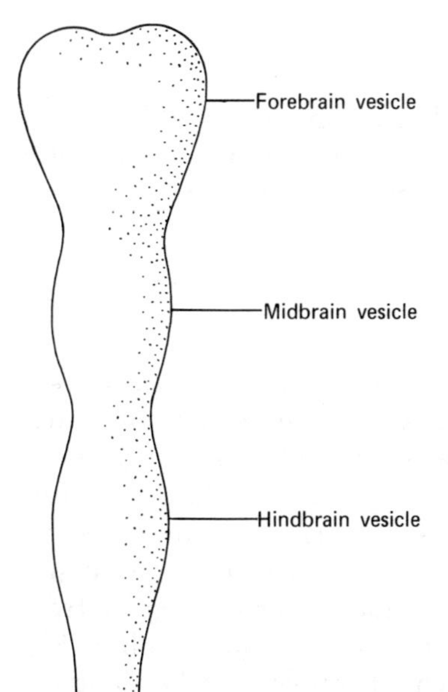

FIGURE 1-4. The embryonic vesicles of the forebrain (prosencephalon), midbrain (mesencephalon), and hindbrain (rhombencephalon).

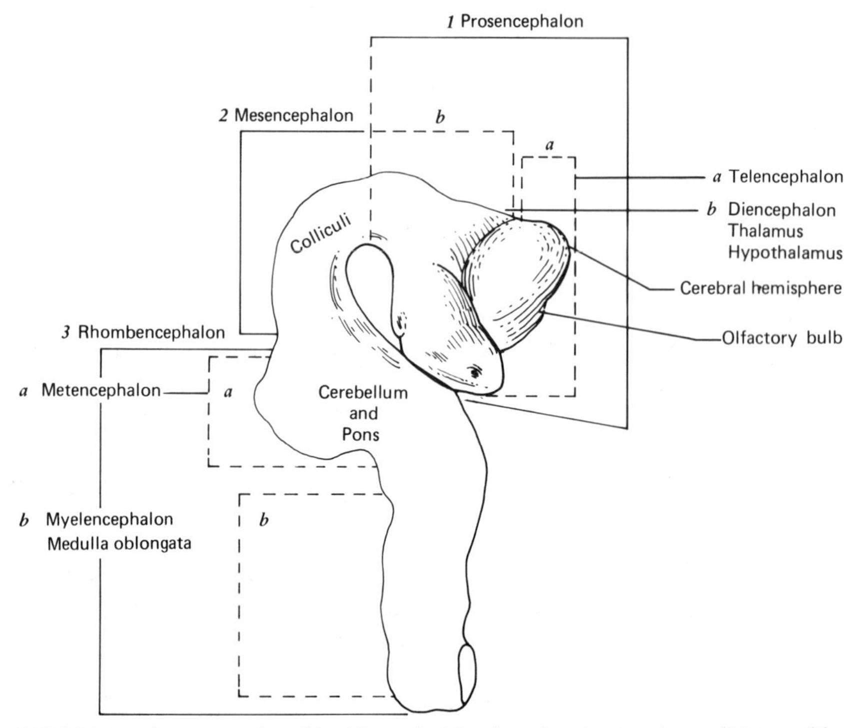

FIGURE 1-5. A representation of the 4½-week-old embryo showing the primary divisions of the brain and the structures that develop from them.

the cerebellum and also contains a number of important vital centers. Finally, the myelencephalon becomes the medulla oblongata, a center for the control of the heart and mechanisms of respiration.

NERVES KNOW THEIR MUSCLES

At an early stage in the development of the nervous system, nerve cells begin to grow out from the spinal cord, eventually to become either sense organs or to develop into neural fibers that control the muscles and glands. Cells from just outside the spinal cord grow toward the receptors to provide connections between the central nervous system and the sense organs. Connectors are also needed between the clusters of cells just outside the cord and centers within, and fibers therefore grow from sensory cell bodies toward the cord to connect to centers within (Fig. 1-6). The final result is the complex system of centers that lie along the

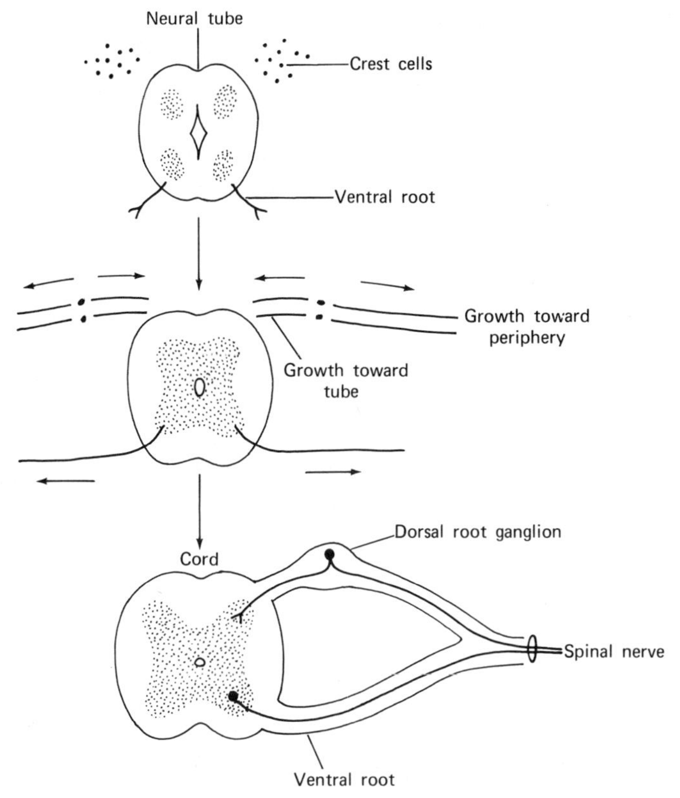

FIGURE 1-6. A schematic diagram of the growth of the dorsal and ventral spinal roots during embryonic development.

axis of the head and spinal column, which is known as the central nervous system, and the massive network of nerves that innervates all parts of the body and is known as the peripheral nervous system.

IT GOES TO THE HEAD: ENCEPHALIZATION AND CORTICALIZATION OF FUNCTION

Implicit in much of the preceding discussion is the principle of encephalization of function, that is, as evolution progresses from lower to higher forms, the functions once exclusively mediated by lower centers are either taken over by the brain or are brought under its domination. In addition, corticalization of function is also the rule when we arrive at the species level where the cortex first appears. Corticalization refers to the principle that the cortex functions as the highest center in the nervous system and dominates the lower centers.

The principle of encephalization of function is evident not only in phylogeny but in ontogenetic development, or the development of the individual, as well. This is seen in the more rapid development of the head region in the fetus and in the observation that the head and upper body in the infant come under control before the legs. This is why infants must go through the long sequence of first holding up the head, then sitting up, crawling, standing with help, and only after a year of postnatal development, on the average, walking unaided. However, we must emphasize again that while encephalization and corticalization are the rule, the older, more primitive parts of the brain remain functional and make their specialized, if diminished, contributions. The cortex, therefore, is a kind of general headquarters of the brain in charge of the lower centers, just as a military headquarters unit commands all of its lower echelons.

A CREATURE OF HABIT: THE ASSOCIATION AREAS

From his own and no doubt biased point of view, man sees himself as sovereign over all the lower forms. There is considerable evidence to support this point of view in the amount of cortical tissue provided for association functions in man as compared with that set aside for the "sensory projection" areas, so-called because the stream of neurons coming in from the receptors "project" out onto the cortex somewhat as a slide projector throws an image on a screen (Fig. 1-7). The motor region of the cortex is also included in the category of projection fibers—in this case, of course, projecting downward. Association areas, on the other hand, function in the so-called higher mental processes of memory, perception, reasoning, and the like.

As Figure 1-8 reveals, most of the rat's cortex is identified with specific sensory and motor functions whereas man's is largely associational in function. We should also note that the frontal regions of the brain containing large masses of associational cortex are very prominent in man. This preponderance of associational tissue in the human species reflects man's life style—that is, he is a creature of habit and (sometimes) of thinking rather than one of reflexes or of simple stimulus-response reactions. And so whether we use it or not, nature has provided us with a brain admirably suited for learning and reasoning.

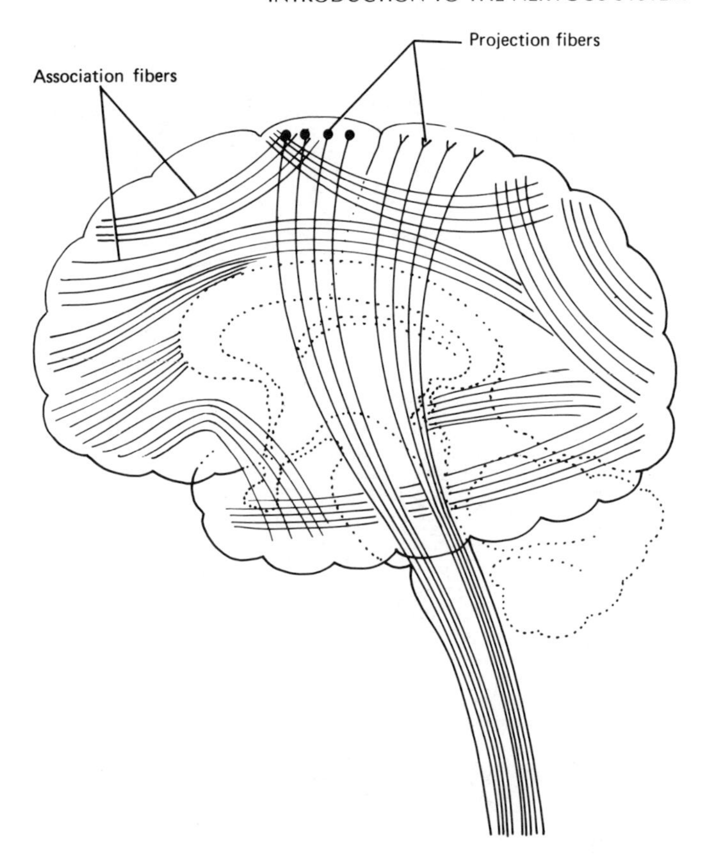

FIGURE 1-7. A schematic diagram of the brain and spinal cord showing projection and association fibers.

AFRICAN GENESIS: A NOTE ON THE BRAIN OF PRIMITIVE MAN

The nostalgic bit of folk wisdom which tells us that bygone times were better than the present is not supported by anthropological and archaeological estimates of the size of the brain of primitive types of man (Fig. 1-9). The reconstructed skulls of several primitive types show that man's cranial capacity was much less than that of *Homo sapiens* (*homo,* man; *sapiens,* wise) the only surviving species of the genus, *Homo.* The illustration also reveals that the frontal part of primitive man's brain was inferior in development to that of modern man. This is especially significant in that the frontal poles of the brain serve higher intellectual

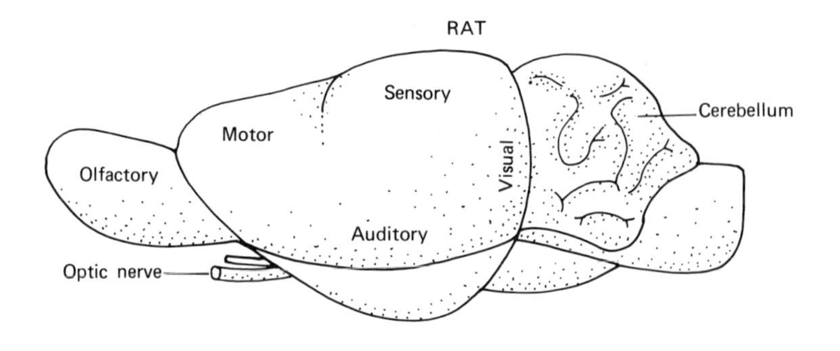

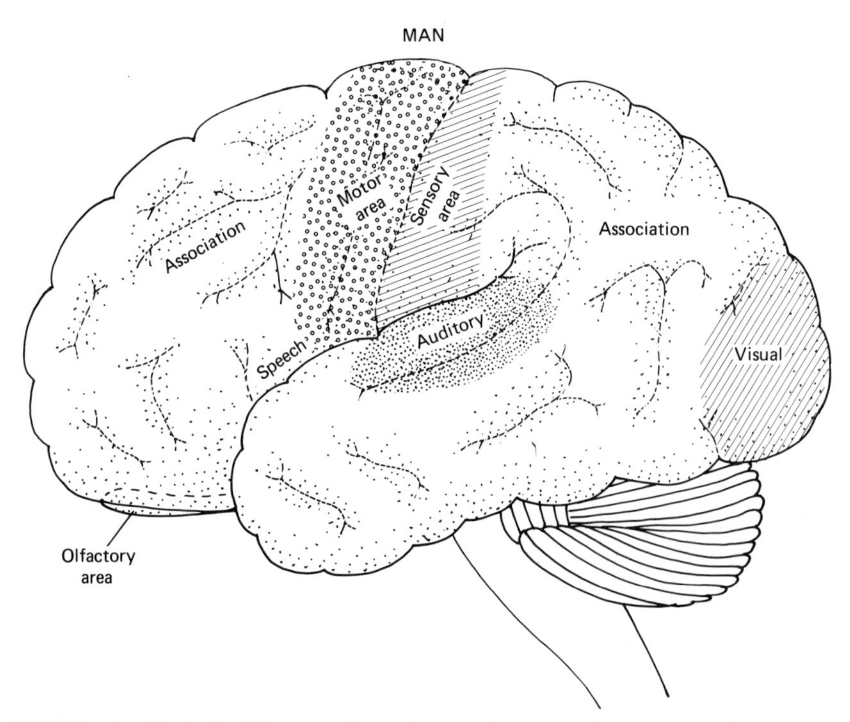

FIGURE 1-8. The brain of the rat and man showing the relative size of the specific motor and sensory areas as compared to association areas. Note that in man the largest component of the cortex is associational.

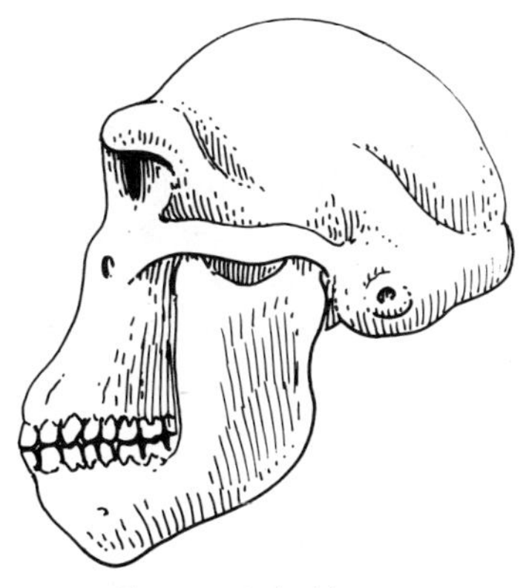

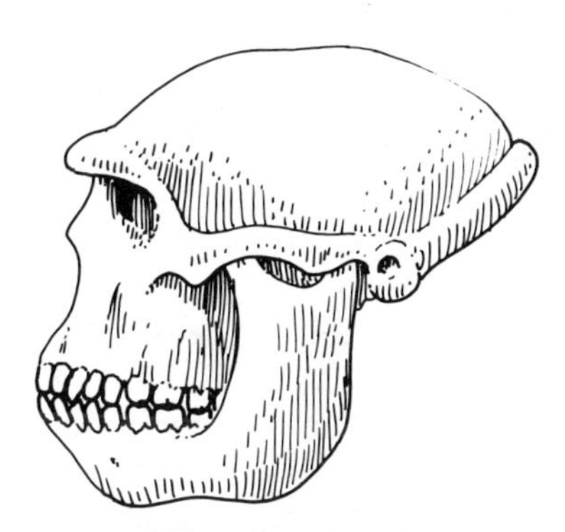

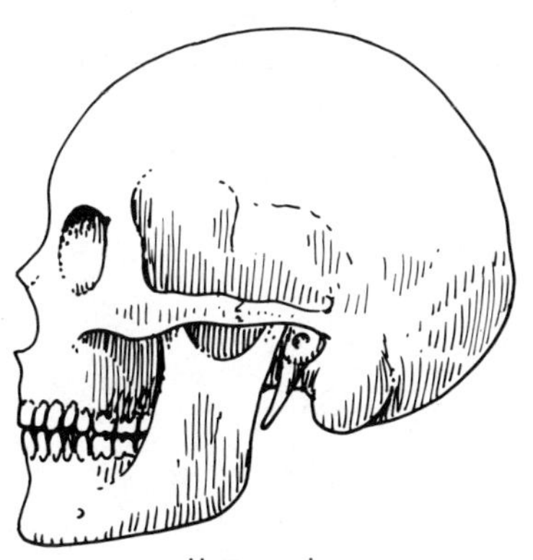

FIGURE 1-9. The skulls of two primitive types of man compared with that of *Homo sapiens*. Note that the brain cavity of modern man is about three times greater than that of *Homo australopithecus*.

functions of judgment, planning, and the management of behavior. Large lesions to this region leave the individual with a serious loss in abstracting ability. Such injuries are also likely to result in a person who lacks control over his impulses and emotions. This suggests that primitive man may have been a rather formidable creature, lacking in the more sophisticated ways of thinking and also somewhat savage in disposition—not a welcome person as a dinner guest.

NOT CREATED EQUAL: THE GENIUS AND THE RETARDED

Some brains develop in such a manner as to become highly complex organs capable of thinking innovatively and creating new forms in the arts and sciences. Others are lacking in some respect and their possessors are retarded—some so severely that they must be cared for by attendants. They cannot learn even simple skills or, in the stately language of the English common law, "how to avoid the common dangers of life."

We do not always know the reasons why one brain is average, another retarded, and a third superior. Presumably both genius and retardation are dependent on how the genetic code directs neural development into more or less complex interconnections. Environment, too, is an important factor. Intelligence cannot flourish in an intellectual and cultural vacuum. Beethoven's and Einstein's great genius in their respective fields was nourished by rich artistic and scientific environments. Any shepherd boy can whistle a tune or count his sheep, but only a genius with environmental advantages can sail on uncharted seas of mathematical reasoning or take a simple folk melody and ennoble it into a Ninth Symphony. All development and all behavior is, therefore, the result of the interaction between heredity and environment and not simply the product of one or the other.

HOW IT IS ALL GOTTEN TOGETHER: METHODS IN NEUROPHYSIOLOGY

There are many methods for studying the nervous system. The most fundamental of these is the oldest of all, the anatomical method, whose tools are skillful dissection and keen observation. Our basic knowledge of neuroanatomy came from the dissecting table, and it is part of our cultural heritage from ancient Greece and Italy. These beginnings ex-

EXPERIMENT I

HELMHOLTZ MEASURES THE SPEED OF THE NERVOUS IMPULSE

In the 1830s the German physiologist and founder of modern experimental physiology, Johannes P. Müller (1801–1858) taught that the velocity of the nervous impulse was so great that no stretch of nerve long enough to use as a test preparation could be found to measure it. He proposed that the speed of the impulse must approach that of the speed of light, so that a nerve of the dimensions of outer space would be required to obtain two test points sufficiently distant from each other to measure it.

Hermann L. F. von Helmholtz (1821–1894), the great German physicist and physiologist and former student of Müller's, measured its speed in 1851 in a simple frog nerve-muscle preparation a few inches in length. First he stimulated the motor nerve of the gastrocnemius muscle of the leg close to the point where it enters the muscle. With the help of a tuning fork whose rate of vibration was known he calculated the time it took for the muscle to react. Next he stimulated the nerve as far from the muscle as possible and again calculated the time for reaction. The difference he took to be the time required for the impulse to travel over the nerve. He found this interval to be about 30 meters per second—rapid, to be sure, but nowhere near the speed of light.

Today's neurophysiologist has far more sophisticated techniques available for measuring neural velocities, but Helmholtz's pioneer experiment provided a foundation for modern studies of the electrophysiology of nerves.

plain why so many terms in neuroanatomy are of Greek and Latin derivation. They are attempts to express what the structure looked like or how it seemed to function to its discoverer. Whoever first sliced through the cerebellum must have been struck by the cedarlike appearance of the white and gray matter, since he named it *arborvitae,* the tree of life. Similarly, the anatomist who first dissected away the covering of nervous tissues was surprised at the toughness of the outer layer, the webbed delicacy of the middle layer, and the tenderness of the inner layer. He

CASE I

THE DISCOVERY OF THE SPEECH CENTER BY BROCA

In 1831 a patient named Leborgue, aged 21 years, was admitted to the Hospice de Bicêtre, a mental hospital near Paris, with a single defect—he was unable to speak, except to say "Tan." He became known throughout the hospital by that name and appeared to be normal except for his aphasia, or loss of the ability to speak. He could, however, communicate intelligently by means of sign language and gestures, and showed no defects of the musculature of the larynx, lips, or tongue.

Solely because of his inability to speak, Leborgue was confined at the Bicêtre for thirty years until 1861 when, having developed gangrene, he was transferred to a general hospital under the care of the surgeon, Paul P. Broca (1824–1880). Broca examined the patient over the five-day period between his admission and death, satisfying himself that there was no organic articulatory defect or paralysis. On the day following his patient's death, Broca performed an autopsy and discovered a lesion about the size of a hen's egg in the third frontal convolution of the left cerebral hemisphere. The tissue showed evidence of degeneration and was adherent to the dura, or covering of the brain. The right side of the brain was normal.

Broca presented the case to the Anthropological Society in Paris expressing his conviction that this area was the speech center. Shortly afterward he presented a second case demonstrating precisely the same type of lesion in the identical, circumscribed area. This center is now known as Broca's area and is recognized as the motor area for the control of speech.

Broca's careful examination prior to his patient's death, and his recognition of the broad implications of the two cases that he presented, form a model of the clinical method. It was the first direct evidence that there is localization of function within the human brain.

called them dura mater (*durus,* hard; *mater,* mother), the arachnoid (*arachonoeides,* weblike) and the pia mater (*pia,* tender), respectively.

The anatomical method is complemented by the experimental method, which employs a variety of techniques such as extirpation or removal of parts, the application of stimulating or recording electrodes, and the administration of drugs. Within the past thirty years, the use of these techniques has brought about an explosion of knowledge about the functions of the nervous system, particularly of the higher centers. The advantage of experimental techniques lies in their precision. The neurophysiologist can control the genetic and environmental histories of his subjects. He is able to measure dosages or to control electrical stimuli carefully. He can destroy precisely circumscribed areas in the brain. By observing the subject's preoperative behavior and comparing it with postoperative behavior, he determines the function of the area which was stimulated or removed. The chief disadvantage of experimental methods is this: they are generally restricted to animal subjects. While we can—and do—generalize from animals to men, there is always the danger that our conclusions may be in error because of the great differences in the complexity of the nervous system of man and animals.

Our knowledge of the functioning of the human nervous system has been largely obtained through the clinical method. Neurologists and neurosurgeons provide us with case histories that reveal the effects of lesions to the nervous system. Such a record will include information about the patient's background, the onset of the disease or accident, and his subsequent behavior. Any corrective measures taken are described along with their behavioral consequences.

The chief disadvantage of the clinical method is that it is nature-made, and we, therefore, have no control over its circumstances or over the site and extent of the lesion. Injuries and malignant tumors in particular leave irregular lesions whose boundaries may be difficult to map. However, as judges often say, one case does not make a law. Many clinical cases must be analyzed carefully and compared before a generalization can be formulated about the function of a center or nervous pathway.

KNOWING OUR WAY AROUND

Because the nervous system is a three-dimensional solid, we need terms to describe locations. When we are dealing with a subject in the standing or upright position, structures toward the head are described as cranial or superior, and those toward the feet as caudal or inferior. If our subject is prone, or if we are dealing with an animal who walks on all fours, structures toward the back are said to be dorsal; those toward the abdomen or belly, ventral. The head region is anterior, the tail posterior. Occasionally the term rostral (*rostrum,* beak) is used to refer to the head region.

A horizontal section across the brain or spinal cord is called a cross section or horizontal section (sometimes abbreviated X-section), and one parallel to the longitudinal plane of an organ is known as a sagittal (*sagitt,* arrow) section.

Lateral refers to the side, central or median to the center. Peripheral is also used to refer to nerves or systems that are not part of the central nervous system, which is defined as the brain and spinal cord. When we want to distinguish between that which is near as opposed to that which is far we use proximal. For far we use distal.

There are many other terms of location that apply to special parts of the nervous system. These will be introduced as they are needed in connection with our description of these systems.

2
THE STRUCTURE AND FUNCTION OF NEURONS

Shedding a tear, blinking an eyelid, driving a car, playing a Chopin étude, or trying to fathom Einstein's theory of relativity—all depend on neurons. In this chapter we will introduce these versatile microscopic structures possessed in such abundance by all of us yet seen by very few. In the first part of the chapter our task will be anatomical—to visualize and describe the structure of neurons. In the second part we will begin the study of their functioning—a task that will occupy us, in a sense, for the rest of this book; for, however complex a neural center may be, fundamentally it is a collection of neurons.

BRICKS FOR A HOUSE OF MANY MANSIONS

The nervous system is large and highly complex, and like any bodily organ system, it is an integrated collection of cells of various types. These cells or building blocks provide organs and organ systems with their distinctive structures and make possible their specialized functions. To understand the nervous system, it is necessary, therefore, to become familiar with the elements out of which it is built.

Nerve cells, like other cells, exhibit certain fundamental living processes. These are irritability, conductivity, and metabolism, the last of which includes the ability to utilize food products delivered by the bloodstream for the production of energy and to discharge waste products. Many bodily cells possess contractile ability or the capacity to shorten. There are masses of these cells in the muscular, digestive, and vascular systems whose job it is to contract for the purpose of moving limbs, pumping blood, or digesting food. Other cells may have secretory functions, producing either hormones that are emptied directly into the bloodstream or glandular products that are discharged outside of the body or into the digestive tract. The sex cells have the specialized function of reproducing themselves in order to provide for the continuity of life in multicellular organisms. Nerve cells are highly specialized in two of these basic life processes, irritability and conductivity, and after we have described the structural properties of neurons, we shall consider these specialized functions in some detail.

Before we describe the anatomical characteristics of neurons, it will be helpful to consider the general structural properties of cells, since neurons share these with other cells. As Figure 2-1 shows, cells are bounded by a membrane that consists of proteins and lipids. The cell membrane is quite thin—about 75–100 Ångström units (hundred-millionths of a centimeter) in thickness. The cell membrane has as one of its functions that of providing a boundary for the cell, but more importantly it acts as a selective gateway allowing some materials to enter the cell even as it prevents others from doing so. Similarly, the membrane

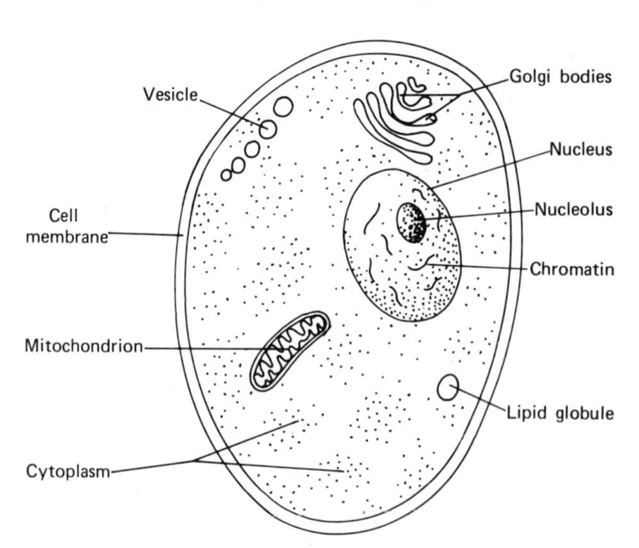

FIGURE 2-1. The internal structure of a typical cell.

selectively allows certain products to leave the cell while keeping others in. Measurements across the cell membrane also show voltage or potential differences. We shall see that this ability to screen products that enter and that leave the cell and to maintain potential differences is of fundamental importance in the transmission of the nervous impulse.

A POWERHOUSE OF ENERGY: CYTOPLASM

The inside of cells is filled with cytoplasm, a complex substance consisting of water, organic compounds—fats, carbohydrates, and proteins—electrolytes (molecules capable of carrying electric currents), enzymes, and organic salts. Also interspersed within the fluid medium of the cytoplasm are threadlike structures, vesicles or sacs, and solid particles that can be stained to appear quite dark against the lighter background of the cytoplasm. Somewhere near the center of the cell is a relatively large rounded body called the nucleus (*nucela,* little nut) with its dark center, the nucleolus, a kind of center within a center. The nucleus has the important responsibility of supervising the metabolic processes going on within the cell. If it is destroyed, the cell will die. The nucleus is also concerned with cell reproduction and contains within itself the genetic material, deoxyribonucleic acid (DNA). The nucleolus is rich in ribonucleic acid or RNA which, during cell division, serves as a messenger carrying instructions from DNA to the protein factories in the cytoplasm whose responsibility it is to provide material for daughter cells. It should be noted, however, that adult neurons do not reproduce themselves. Some are capable of self-repair following injury, but unlike adult skin cells or sex cells that can reproduce, nerve cells cannot. This, we shall find, has important implications for limiting the recovery of function following disease or injury to the nervous system.

Among the more important substances that can be observed in the cytoplasm around the nucleus are first, the Golgi complex (named after Camillo Golgi, 1843–1926, an Italian anatomist) whose functions are not completely understood but which are believed to be associated with secretory functions in secretory cells and with enzymatic activities that take place following injury to the neuron. The Nissl substance (named after Franz Nissl, 1860–1919, a German neurophysiologist) consists of complex masses of tubes, vesicles, and deposits of RNA whose functions are associated with protein synthesis. Lending support to this hypothesis is the finding that the Nissl substance is depleted by extreme neuronal fatigue and that it also shows changes during the regeneration of a neuron following injury. Some recent research suggests that RNA may also be involved in neural changes that are assumed to occur as a result of learning.

The mitochondria are the powerhouses of the cell, for it is here that the oxidation of various food substances takes place. The mitochondria can extract energy from nutrients and oxygen and store that energy in the form of phosphate bonds (adenosine triphosphate or ATP) for quick release during neural transmission.

Other substances that may be found in cells are lipids or fat deposits, secretory granules, melanin pigment, networks for the metabolism of proteins, and lysosomes capable of digesting unwanted substances within the cell. We need not be concerned with the details of these cytoplasmic constituents here.

NEURONS: MANY SHAPES, MANY SIZES

Neurons come in all sorts of shapes and sizes depending on their placement in the nervous system and the functions for which they are responsible. Figure 2-2 shows some of the more important varieties diagrammatically. Here we shall be concerned with three basic types, sensory, motor, and internuncial neurons. These three kinds of neurons are the means by which impulses are conducted toward and away from the central nervous system and are transmitted from center to center within that system.

The typical motor neuron consists of a cell body or soma, a number of dendrites, a long axon and a terminal branching called a telodendron (*tele,* distant; *dendron,* tree). It may also give off collateral branches (Fig. 2-3). The axon, in turn, consists of an axis cylinder, a myelin sheath, and a layer of cells called the neurilemma or Schwann cells (named after Theodor Schwann, 1810–1822, a German physician). The functions of each of these structures will be discussed next.

The cell body of motor neurons lies within the central nervous system and governs the metabolism of the cell. Its structural and functional well-being is essential to the life of the cell. For generations before the development of the Salk vaccine for poliomyelitis, thousands of children and many adults each year became the victims of this viral disease that kills the cell bodies of the motor neurons that originate within the spinal cord and control the voluntary muscles. Many victims were crippled as an aftermath of the disease. Some in whom the virus destroyed the cell bodies of the neurons in the respiratory centers in the medulla died as a result of respiratory failure.

Dendrites (*dendron,* tree), whatever their shape or size, carry im-

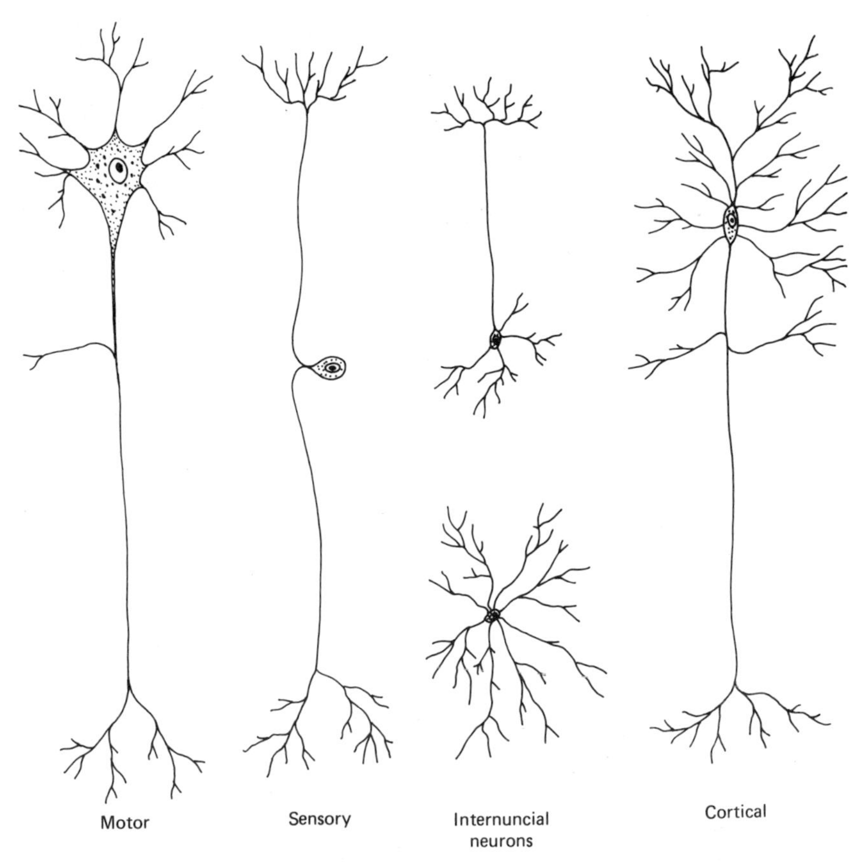

FIGURE 2-2. Types of neurons.

pulses toward the cell body. They are the collecting or receiving agents for the cell, picking up and transmitting stimulation from adjacent cells.

The axon (*axon,* axis) is the long filament—sometimes two to two and one-half feet in length—which originates at the axon hillock and terminates in a muscle or gland at the teledendria where the neuron comes into functional contact with muscle fibers or the secretory cells of glands. The axon's function is to conduct impulses away from the cell body. Therefore, to summarize neuronal transmission, it is from dendrite → cell body → axon → telodendria → muscle or gland.

Many axons are covered with a fatty layer called the myelin sheath (*myelos,* marrow), which is laid down so that it forms regular indentations along its length known as nodes of Ranvier (named after Louis A. Ranvier, 1835–1922, a French histologist). This sheath gives the axon an

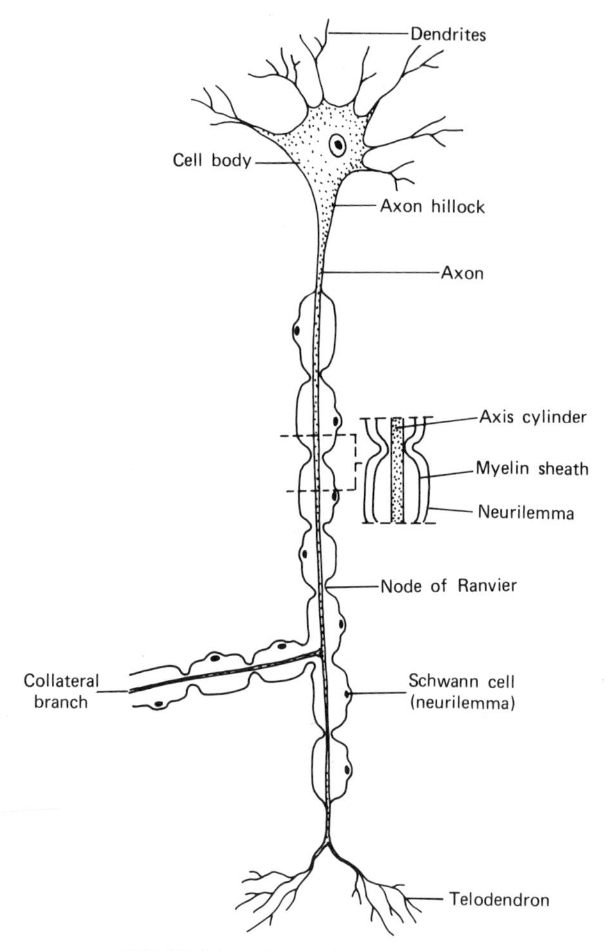

FIGURE 2-3. A typical motor neuron.

appearance not unlike that of a string of sausages. Myelin sheaths are found typically around large neurons, which conduct more rapidly than smaller, unmyelinated neurons. Myelin thus has an insulating and facilitating function in neuronal transmission.

The neurilemma (*neuron,* nerve; *lemma,* skin) is found only around axons located outside the central nervous system, such as those that conduct sensory and motor impulses to and from the periphery of the body. The neurilemma consists of a sheath of cellular material called Schwann cells. The function of the neurilemma is to lay down or form the myelin sheath. The neurilemma also provides a channel for axonal regeneration if the axon is cut or injured. Because of the special clinical importance of regeneration, we shall discuss this process in more detail later in the chapter. Outside the neurilemma is a layer of collagenous material that provides a structural framework or sheath for the axon.

Sensory neurons have all the same structures as motor neurons, but the physical arrangement of the parts is different, reflecting the different functions of the two types (Fig. 2-4). Note that the sensory neuron is T-shaped with the cell body lying just outside the spinal cord in an enlargement of the sensory nerve trunk just before it enters the cord.

The dendritic process conducts impulses to the cell body from a receptor and into the spinal cord where its telodendria link up with other neurons conveying impulses upward toward the brain or where it may connect with motor neurons involved in reflex arcs. Note that the long dendrite of the sensory neuron is structurally identical with the axon. It is classified as a dendrite because it conducts impulses toward the cell body.

As Figure 2-4 shows, the elongated dendritic and axonal processes of sensory neurons are covered with myelin and neurilemma.

Internuncial neurons are of many sizes and shapes depending upon the site they occupy in the nervous system and the function they serve. Some are relatively long, such as those that travel up and down the cord; others are short, like those that link one region of the cerebral cortex to another. Many of these cortical neurons are incredibly complex (Fig. 2-2), branching and proliferating intricately.

In general, the internuncial neurons that make up conduction pathways in the spinal cord are myelinated. However, as was pointed out earlier, neurons within the central nervous system do not possess neurilemma and are thus incapable of regeneration.

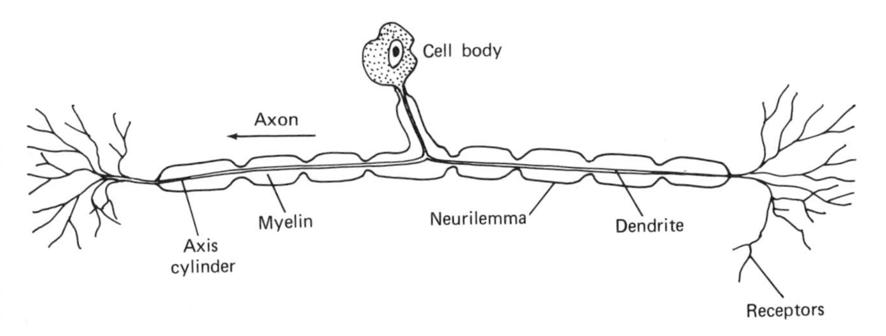

FIGURE 2-4. A typical sensory neuron.

MORTAR FOR THE BUILDING: GLIA CELLS

It may come as a surprise that the most numerous cells in the nervous system are not neurons but neuroglia or glia (*glia,* glue) cells, as they are commonly called. Generations of neuroanatomists regarded them as supportive tissues without any positive functions, but recent studies indicate that they may have two other equally important roles to play. Some glia cells act as phagocytes (*phagos,* eating; *kytos,* cell) or cells that, like white blood cells, have the ability to ingest foreign matter or to scavenge dead bits of neurons that may be present as the aftermath of injury or disease. It is also believed that, because of their intimate association with the capillaries of the blood vessels that serve neural tissue, at least some glia cells may have metabolic functions. This is particularly true of astrocytes, star-shaped neuroglial cells whose processes appear to be in close contact with capillaries. Astrocytes may in fact act as transport systems for food products carried from the capillaries to the cytoplasm of neurons.

We should also note that unlike neurons, glia cells are capable of dividing by mitosis and so producing daughter cells. This characteristic is of particular importance in that most of the tumors originating in the brain are the result of the pathological division of glia cells. Because of their rapid rate of division and the attendant pressure created within the closed confines of the skull by the developing mass, neuroglial growths are often fatal (see Case X).

COPING WITH DISASTER: DEGENERATION AND REGENERATION IN NEURONS

In the evolution of the skull and vertebral column, nature provided the central nervous system with an effective suit of armor. The peripheral nervous system, however, did not fare so well, exposed as it is to the dangers of cuts and bruises. However, even the thickest skull or stiffest back cannot prevent bacteriological and viral infections from finding their way within. And, unfortunately, man's brain has all too often been employed in devising weapons and other implements that can penetrate even the strongest armor.

While evolution has not yet found a way to protect us against our own folly and, in fact, is not always successful in coping with its own forms of destruction, it has developed a remarkable mechanism for restoring function following damage to the nervous system. Sometimes this restoration of function involves making do with what is left and learning how to adapt to living with limited function. We shall have more to say about adapting and relearning when we discuss the brain. It may also involve, however, the actual regeneration of neurons. This renewal can happen under two conditions: (1) the neuron must not be too severely injured (the cell body must be intact), and (2) the neuron must possess neurilemma. Since only neurons of the peripheral nervous system are equipped with neurilemma, it follows that only they can regenerate following injury. When the destruction occurs in the central nervous system, function may be lost. If partial or even complete recovery occurs following central neuronal destruction, it must be through the use of alternate pathways or centers.

When axons are severed in the peripheral nervous system, a process known as Wallerian degeneration (named after Augustus V. Waller, 1816–1870, an English physician) takes place. This process involves changes in the cell body and in the portion of the axon distal to the injury. This is shown diagrammatically in Figure 2-5. As a study of Figure 2-5 reveals, the portion of the axon distal to the cut degenerates

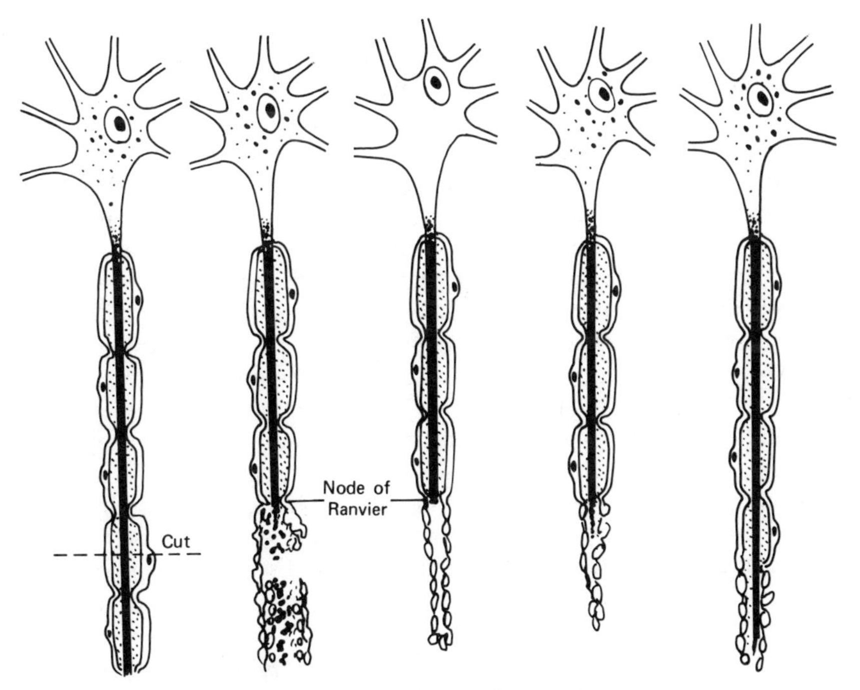

FIGURE 2-5. Wallerian degeneration.

CASE II

MALT AND McKHANN REJOIN A SEVERED ARM

In the past fifteen years the replantation or rejoining of severed limbs has become a relatively commonplace operation at large medical centers. In the spring of 1962 when a twelve-year-old boy was brought to the Massachusetts General Hospital thirty minutes after his right arm had been sheared off at the shoulder between the side of a freight train on which he was hitchhiking and a stone abutment, the procedure was comparatively rare. Because the boy's general condition was excellent and the severed limb in good condition below the level of the injury, it was decided by surgeons Ronald A. Malt and Charles F. McKhann to attempt to rejoin the arm.

The severed limb was taken to the operating room, perfused with Ringer's solution, heparin, and antibiotics to clear any remaining blood, arrest cellular deterioration, and prevent possible infection. The limb was x-rayed and an arteriograph taken to assure the surgeons that the bone structure and circulation were intact.

After the humerus had been rejoined by a special pin to the shoulder stump, the veins and arteries were reconnected and circulation was restored. The ends of each of the four major nerves of the arm were joined with a marking stitch for future identification. Muscle tissues were rejoined and skin grafts applied over the denuded area. A cast was placed on the limb and a program of daily muscle exercise initiated by application of galvanic shocks in order to prevent muscular degeneration from disuse until nerve regeneration had occurred.

In September 1962 a second operation was performed to join the nerves. It was found necessary to clear away neuromas (scarlike tissues) at the ends of the nerves and to join two of the nerves with grafts taken from another site in the patient's body in order to close gaps.

Four months later contraction appeared in the flexor muscles of the forearm, whose strength returned rapidly. Within eleven months the two-point discrimination was present along with sensations of temperature and touch, indicating that both motor and sensory neuronal regeneration was taking place. Twenty months after the second operation strength in the biceps and opposing extensor muscles was between 30 and 40 percent of normal. The patient could lift his arm at the shoulder about 75 degrees. The elbow had recovered 90 percent of full extension, and the finger and wrist muscles and joints had recovered 80 percent of capacity. Subsequent operations were planned to correct tencion deficiencies and strengthen the fracture site. The patient could write his name and lift weights with the reunited arm. However, his boyhood dreams of someday becoming a big league baseball player were doomed, even though he had regained use of his arm to an astonishing degree considering the nature of his injury—a condition to be fully appreciated when one considers the alternative of going through life with a missing arm or crude hook prosthesis.

Summarized from Ronald A. Malt and Charles F. McKhann, "Replantation of severed arms," *Journal of the American Medical Association,* September 1964.

as it does toward the cell body; however, in the latter case only back to the first node of Ranvier. In the parts of the axon where degeneration is going on, the myelin sheath and axis cylinder break down and are either absorbed or phagocytized by invading macrophages. Back in the cell body the process of chromatolysis, or gradual disappearance of the Nissl substance, indicates that the cell's headquarters is also affected by the injury and is responding by undergoing metabolic changes. The enlargement of the nucleus and nucleolus and their displacement toward the side of the cell body further indicate that peripheral injury results in profound changes in the cytoplasmic processes within the cell and probably involves changes in protein metabolism necessary for repair. As has already been pointed out, the neurilemma persists after injury, forming a kind of pathway along which the fibrils of the axis cylinder slowly begin to grow toward the point of termination. Growth proceeds at the rate of about 2–3 millimeters per day until the telodendria are reached and function is once again restored. During the process of axonal regeneration, the myelin sheath is rebuilt.

If trauma or injury is too severe with proliferation of scar tissue, the neurilemma may be destroyed or rendered impenetrable by the fibers of the axis cylinder, and in this case function will be permanently lost. In those dramatic instances where severed limbs have been surgically replaced, the degree of restoration of function depends on how clean the original separation was, how the limb was handled in transit (keeping the limb cool is important) and, of course, on the skill of the surgeon (see Case II).

EXCITATION AND CONDUCTION IN NEURONS: SWIFT MESSENGERS

If there were such a thing as a neuron long enough to reach across the United States, a nervous impulse could travel from Boston to Los Angeles in approximately ten hours. In your automobile, driving reasonably, it would take a week. This remarkable ability of neurons to conduct impulses swiftly depends upon a series of complex electrochemical processes that occur millions of times every second throughout our bodies. During the course of these events neurons utilize food, consume oxygen, and produce heat and waste products that must be carried away by the bloodstream. That all of this takes place in time intervals measured in thousandths of a second is eloquent testimony to the wonders of the nervous system.

Theoretically, the neurophysiologist can measure any aspect of neuronal activity—electrical, chemical, thermal—or the volume of waste products resulting from neural transmission. All of these have, in fact, been measured. However, the most conspicuous of the various activities exhibited by neurons are electrical in nature. Moreover, with the development of high fidelity amplifying and recording equipment, the electrical signs of nervous activity are the easiest to measure reliably and precisely.

In order to visualize the nervous impulse, the neurophysiologist typically employs an oscilloscope, which is a kind of scientist's television set that has the capacity to transform electrical energy into visual traces on a phosphor screen. This is what a home television receiver does but in a more elaborate manner. If the neuroscientist wishes to make a permanent recording of neuronal activity, he can take photographs of the oscilloscope screen, or he may employ pen-writing equipment, for example, the electroencephalograph in recording brain waves.

Electrical phenomena are measured and recorded as changes in potential. Potential refers to the amount of electrical charge measured at some point as compared with the amount at some other point. Thus, if we take a small flashlight battery and connect its terminals to recording equipment (Fig. 2-6), a potential will show in the form of a voltage reading on the voltmeter. Voltage indicates the amount of electrical force flowing through a circuit. A small battery will show potentials of 1.5 volts. A neuron can develop a potential difference of 50–100 millivolts or thousandths of a volt—not very much, to be sure, but enough.

The electrical potential present in neurons may be measured by attaching one electrode of a recording device to the outside of the cell

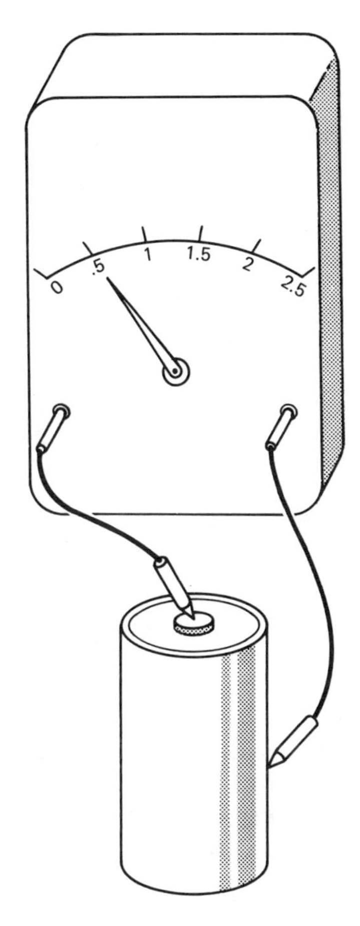

FIGURE 2-6. Arrangement showing a potential change in a voltmeter connected to a battery.

membrane and inserting the other into the axis cylinder so as to record from inside the cell. In this way the potential difference being measured is that which exists across the cell membrane (Fig. 2-7).

THE MESSENGER AT READY: THE RESTING POTENTIAL

The electrical potential generated by the neuron is a result of the fact that the cell, like a battery, is in a state of polarization (Fig. 2-8). In a battery the center terminal is positive while the outside or can of the battery is negative. In neurons the inside of the cell has a relatively high

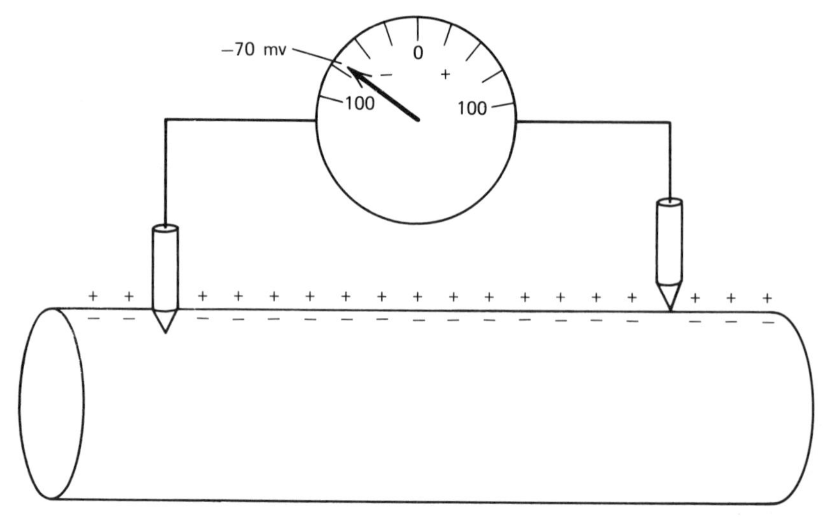

FIGURE 2-7. Method for measuring the potential difference across the cell membrane.

concentration of organic anions. Ions are atoms that have lost or gained one or more electrons and so are left with either a positive or negative charge. Anions have a negative charge. Inside the membrane are proteins, which are the source of the negative charges, anions (A−). Because the cell membrane is impermeable to these anions, they remain inside the cell creating a negative potential. The cell membrane is permeable to potassium ions, (K+) but much less permeable to sodium ions (Na+). The cell membrane tends to transport actively or to pump potassium ions into the cell from the extracellular fluid environment. These tend to

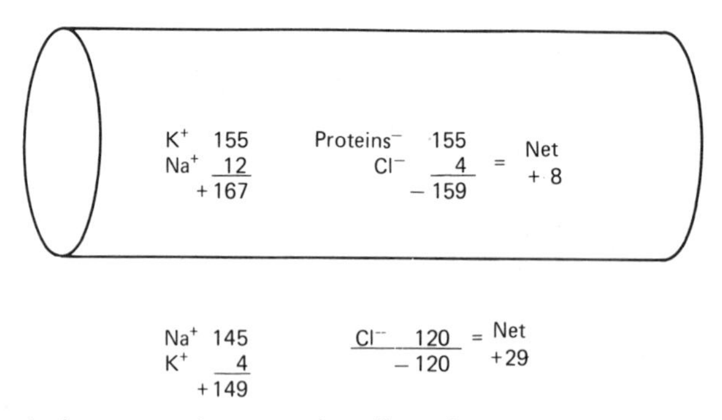

FIGURE 2-8. Ionic concentration across the cell membrane. Note that the interior of the cell shows a higher concentration of negative ions than the exterior.

diffuse back out. However, sodium ions are actively transported or pumped out through the cell membrane, which is much less permeable to sodium ions than potassium ions. Thus, there are many more positive ions outside of the membrane than there are inside. Because the outside of the cell membrane has an excess of sodium ions (Na+) and because sodium ions also cannot move freely across the cell membrane, the outside of the membrane is positive relative to the inside. This potential difference across the cell membrane is known as a resting potential and is on the order of 50–100 millivolts, depending on the neuron being measured and on the extracellular conditions surrounding the neuron. Actually, the situation is much more complicated than our simplified account suggests. There are other ionic exchanges involved (chlorine). We are emphasing the potassium and sodium ions since these are the most important in the development of the membrane potentials. Put as simply as possible, the various ions involved come to a kind of steady state, otherwise if the concentration of negative ions on the inside and positive ions on the outside of the cell membrane built up and up, a potential difference of thousands of volts could develop and the neuron would electrocute itself.

THE MESSENGER AT SET: SUBTHRESHOLD DEPOLARIZING POTENTIALS

The resting potential of a neuron can be quickly and drastically changed by a stimulus (*stimulare,* to goad). A stimulus is any factor that is of sufficient intensity to irritate tissue. It may be mechanical, thermal, chemical, or electrical. In studying potential changes in neurons, experimenters use electrical stimulation that can be precisely regulated to produce changes ranging from a bare minimum to a maximum. Upon the application of a stimulus that modifies the equilibrium of the cell minimally, only a temporary, subthreshold disturbance occurs. This is called a subthreshold depolarizing potential. However, if a sufficiently strong stimulus (threshold stimulus) is applied, a dramatic event occurs. The action potential appears.

THE MESSENGER AT GO: THE ACTION POTENTIAL

The potential change that occurs upon the application of a threshold stimulus is called the action potential because it produces a sharp peak (the spike potential) on the oscilloscope screen. It is on the order of 100 millivolts. First the recording equipment shows a rapid change from −50 to −70 millivolts to zero volts and then moves toward the positive side to about +30 millivolts. All of this takes place within 1–3 milliseconds.

What causes these dramatic changes in potential when the neuron is stimulated with an above-threshold stimulus? The research of neurophysiologists has demonstrated that depolarization of the cell membrane takes place at the point of stimulation. That is, a threshold stimulus affects the cell membrane in such a manner as to change its permeability to the ions present in the extracellular fluid surrounding the membrane and to those that are found within the membrane itself. First of all, sodium ions rush into the cell through the membrane. Remember they are positively charged (Na+) while the inside of the membrane is negative. The sudden presence of a mass of positive ions inside the membrane quickly reduces the negative potential to zero. The neuron is now depolarized. However, depolarization overshoots, so to speak, and an excess of sodium ions builds up inside the cell so that the cell momentarily shows a positive potential. At the same time, there is an accelerated movement of potassium ions to the outside of the membrane, allowing a large quantity of potassium ions to diffuse out of the cell. This diffusion allows the inside of the membrane to return to a negative resting level as the sodium ions are being actively transported or pumped to the outside of the cell. Figure 2-9 shows in simplified form the ionic changes that take place across the cell membrane with a threshold stimulus.

So far, we have been speaking as if the potential changes were confined to the site of stimulation. This is not the case, however. The disturbance is rapidly transmitted over the entire length of the neuron. This is the nervous impulse.

ALL OR NOTHING AT ALL

During the very brief period that the action potential is passing over a given spot along an axon, the tissue is incapable of being stimulated further. Technically, it is said to be in an absolute refractory state. The 1–3 millisecond period during which this is taking place is called the absolute refractory period. It follows logically from these events that neurons behave according to an all-or-none law. By "all or nothing at all" we mean that, provided the neuron does respond with an action potential, then that potential will be as large as the size and the condition of the tissue permit. An analogy is often drawn between the behavior of

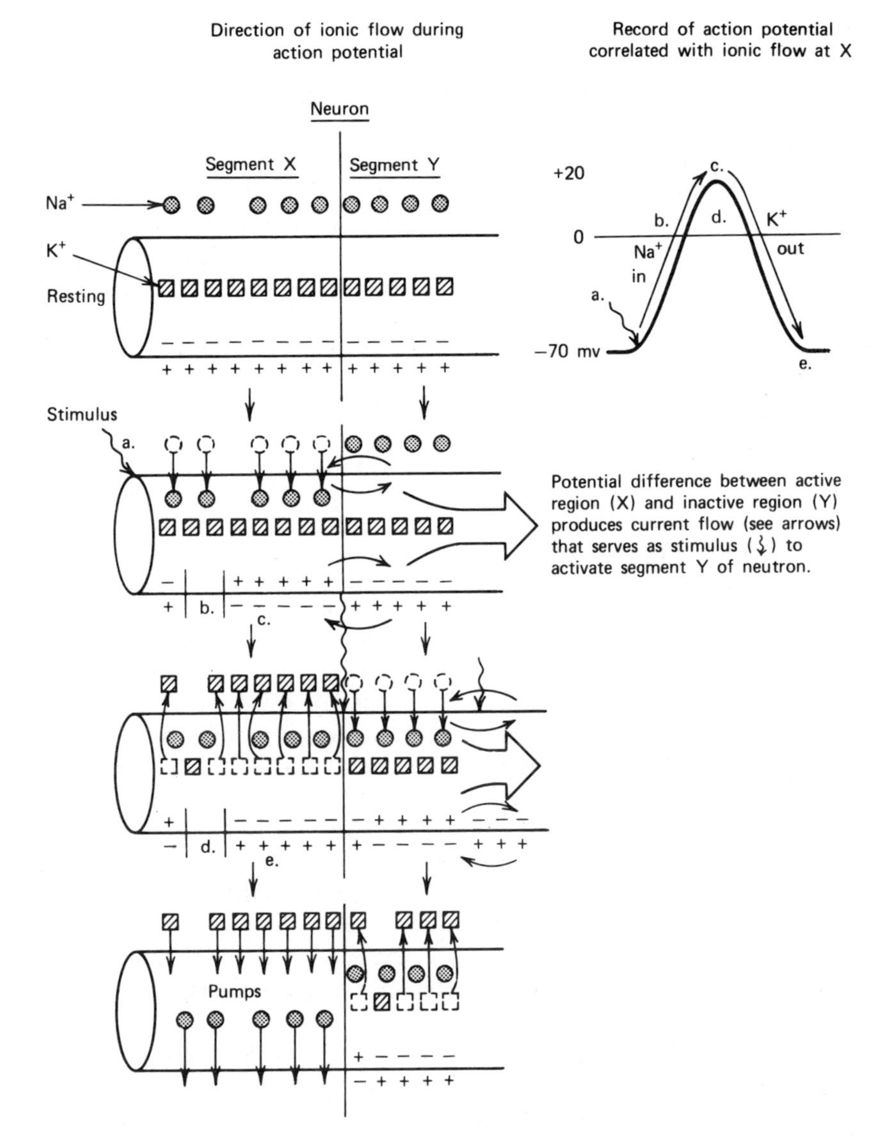

FIGURE 2-9. Direction of ionic flow during the action potential correlated with a record of potential changes.

neurons during the passage of the nervous impulse and the igniting and burning of a trail of gunpowder. Once the trail is "stimulated" (ignited) it rapidly and progressively burns over its entire length leaving no residual powder that can be reignited.

While the all-or-none law is an important generalization about the behavior of action potentials under certain maximum conditions, it does not mean that there are no graded potentials in neuronal tissue. Indeed, we have already pointed out how a subthreshold stimulus produces only a subthreshold potential change that is not propagated. However, if two or more subthreshold stimuli are delivered in rapid succession, they may summate and generate an action potential. It is believed that much of our neural activity may depend upon the summation of graded potentials, and we shall have more to say about these when we consider transmission between two neurons.

REST AND REHABILITATION

Returning to the events of the nervous impulse, it should be noted (Fig. 2-10) that following the absolute refractory phase of transmission there is a relative refractory period when the neuron can be stimulated but only by a stronger than normal stimulus. A stronger than normal stimulus is needed because the membrane is still recovering from the changes that took place during the passage of the action potential.

Since the transmission of the nervous impulse changes ionic concentrations across the cell membrane, these must be re-established. This is accomplished by the sodium and potassium pumps. The sodium pump transports the excess of $Na+$ ions from the inside of the membrane to the outside, and the potassium pump transports the potassium ions that have diffused to the exterior back to the interior. When these events have re-established the normal ionic balance, the neuron has returned to its resting potential (see Fig. 2-9).

The speed of neuronal conduction and the time needed for recovery depend upon the diameter of the axon and the presence or absence of the myelin sheath. In some animals large myelinated neurons can conduct at speeds up to 120 meters per second. The duration of the action potential in these neurons is about .5 milliseconds. In human beings the rate in myelinated fibers is 60–70 meters per second. In unmyelinated fibers the rate may be as low as 1–2 meters per second with action potentials of 2 milliseconds.

In the relatively large fibers of myelinated nerves, a special form of rapid conduction takes place known as saltatory conduction. It will be

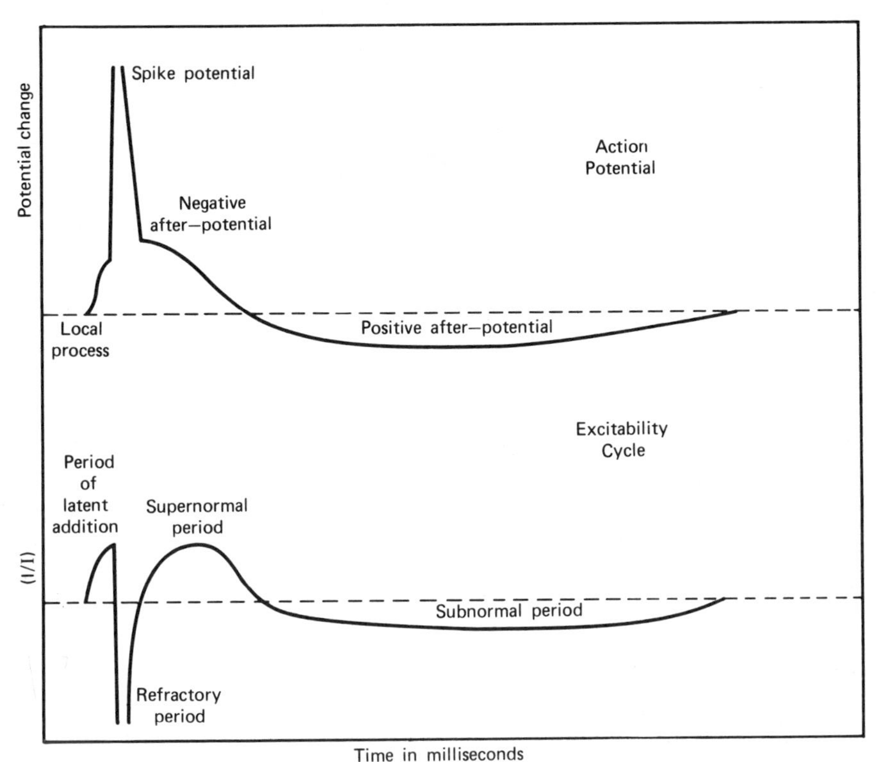

FIGURE 2-10. The composite action potential and excitability cycle of neurons. Reproduced by permission from *Physiological Psychology* by C. T. Morgan. Copyright 1965. McGraw-Hill Book Company, Inc.

recalled that the myelin is laid down in such a manner as to form thin nodes at intervals, the nodes of Ranvier. Because ions cannot flow readily through the thick myelin, impulses tend to conduct from node to node by flowing through the extracellular fluids and axis cylinder. Therefore, in a manner of speaking, the impulse "jumps" from node to node greatly increasing the speed at which it can travel (Fig. 2-11).

The excitability of neurons depends on factors in the cell's environment that either increase or decrease permeability of the cell membrane. Certain drugs or chemical substances can increase excitation through changes in permeability. For example, a low concentration of calcium ions in the extracellular fluids greatly increases the permeability of the cell membrane to the passage of sodium ions. The result may be spontaneous firing of neurons with consequent muscle spasms. This condition is called tetany. On the other hand, local anesthetics, such as pro-

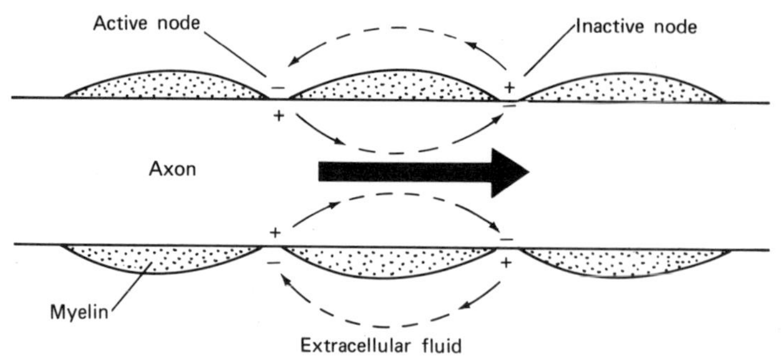

FIGURE 2-11. Saltatory conduction in myelinated fibers.

EXPERIMENT II

THE DISCOVERY OF FIBER TYPES

We have seen how neurons differ in size, shape, and structure, whether or not they have myelin and neurilemma. Sometimes we find fibers of one type of neuron grouped together, as for example in the spinal tracts that conduct impulses up and down the cord. On the periphery of the body, however, large nerves such as the median in the arm or the sciatic in the leg contain a mixture of fiber types. Some of these are sensory in function, some motor, and some are specialized neurons that control visceral activities. The large motor neurons control the voluntary muscles of the arms and legs. The sensory neurons conduct impulses from the skin and from receptors or sense organs in the muscles, tendons, and joints to provide the central nervous system with input from the limbs. The visceral fibers are there to control sweating, the diameter of blood vessels, and the pilomotor response (the hair "standing on end" reaction of shivering or fright).

To discover the action potential characteristics of mixed nerves, early researchers in neuronal physiology attached one electrode to a nerve and the second (the ground or neutral electrode) to another place on the body and then stimulated the nerve. Under these conditions the resulting action potential was found to be a compound made up of different components from different fiber types. Joseph Erlanger and Herbert Gasser, Nobel Prize winning neurophysiologists, in their classic, *Electrical Signs of Nervous Activity* (Philadelphia, 1937), report many experiments on compound action potentials (see illustration).

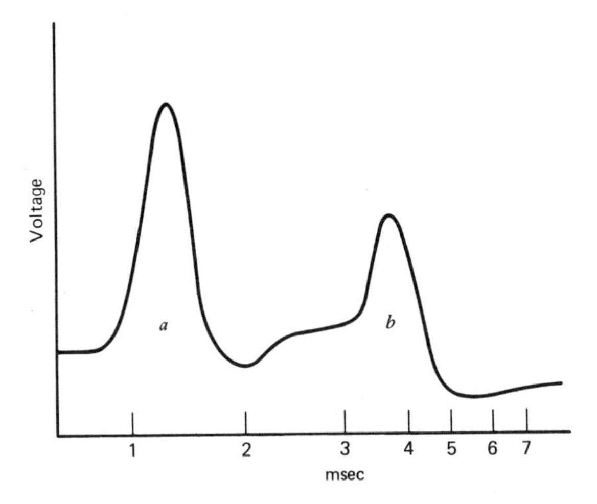

One simple classification of fiber types based on their research is a threefold one designated as *A, B,* and *C* types. *A* fibers are large, high speed, heavily myelinated fibers of the peripheral nerves, and they conduct sensory and motor impulses. The *B* fibers are medium in size, relatively slow in velocity of conduction, and only thinly myelinated. They are found in the connections between the central nervous system and visceral control centers. They are the first links in the chain of neurons that control visceral activity. The *C* fibers are small, unmyelinated, and slow-acting fibers that are found in peripheral nerves. They originate in centers for the control of visceral activity and are the motor neurons that control sweating, blood vessel diameter, and other visceral activities. Only *A* and *B* types were recorded above.

caine or cocaine, decrease permeability to sodium and consequently decrease excitability—a valuable discovery for blocking the transmission of pain impulses during surgery or tooth extraction.

We should also note that because nerves are bundles of neurons carrying thousands of individual fibers, graded rates of transmission are possible, as are differences in the intensity of transmission. This is true because (1) some neurons can fire while others rest; (2) more neurons can transmit more rapidly when stimuli are strong; (3) some neurons have lower thresholds of excitability; and (4) some have larger-sized fibers with thick coverings of myelin. It is for these reasons that we can appreciate differences in the strength of a candle flame as compared with the sun, or the pain of a pin prick as compared with that from a throbbing toothache.

BRIDGES TO CROSS: SYNAPSES

Neurons do not, of course, work in isolation. Impulses must travel over long chains of neurons to get from one center in the nervous system to another. A place on the chain where one neuron joins another is called a synapse (*syn,* together; *hapsis,* joining). The problem of how nervous impulses cross the synapse has been a subject of intense investigation for many years. It is a problem because neurons do not grow together anatomically so that an electrical disturbance can cross directly, but instead neurons are separate structures and thus need some sort of indirect mechanism for the crossing of impulses. We know that neurons are separate entities because studies of the synapse with the electron microscope reveal a minute space between the fibers of the telodendria of one neuron and the dendrites or soma of another.

Traditionally, there were two theories of how the synaptic bridge is crossed, one electrical and the other chemical. The proponents of the electrical theory of nervous transmission were fondly known among their neurophysiologist colleagues as "spark men" and those who supported the chemical theory as "soup men." It turns out the "soup men" are correct. The crossing of the synapse is a chemical process. The terminal portions of the endbrush contain vesicles filled with transmitter substances that are complex organic compounds, such as acetylcholine and norepinephrine (Fig. 2-12). When the impulse reaches the endbrush, these vesicles release their products across the synaptic space onto the dendrites of the adjacent neuron. There they depolarize the dendritic tissues and thus generate an action potential to carry the message onward to its destination.

The question immediately arises as to what prevents the neurons that are stimulated by vesicles from repeatedly firing and thus making any kind of single response impossible. The answer is that substances have been found in the nervous system that are capable of reducing the duration of action of the transmitters and therefore their effectiveness. For example, acetylcholinesterase blocks the action of acetylcholine by breaking it down through enzymatic action and in this way prevents its indefinite accumulation in the synaptic space. Synapses employing norepinephrine, serotonin, and dopamine have also been found in the nervous system. Each of these must have enzymatic or other mechanisms

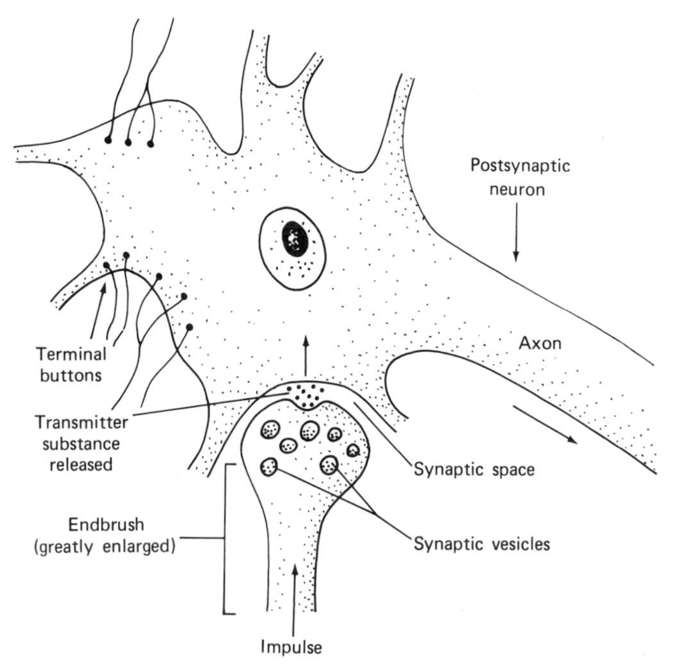

FIGURE 2-12. The structure of the synapse.

for the rapid termination of its action at synapses. These, along with other possible neurotransmitters, will be discussed in chapters to follow.

TOLL GATES AT THE BRIDGE: CROSSING, DELAY, AND INHIBITION

The nature of the synapse provides an opportunity for the selective transmission of nervous impulses. Some are allowed to pass; others fail to get across as threshold stimuli and die out. Moreover, some transmitters at the synapse are excitatory, others are inhibitory. Several of these synaptic processes are of sufficient importance to merit brief discussion.

First we want to emphasize that the synaptic bridge is a one-way street. Impulses do not cross backward from dendrite to telodendria but only from the terminals on endbrushes across to the adjacent dendrites. A moment's reflection about the arrangement of transmitter vesicles will show why this is so. These are located in the telodendria and discharge their chemicals onto the dendrites or soma of the adjacent neuron. Dendrites do not possess such vesicles and so cannot send chemical transmitters backward across the synapse.

Transmission occurs at the synapse when a transmitter substance generates a series of excitatory postsynaptic potentials (EPSPs) in the dendrites of the neuron undergoing bombardment by the vesicles of the transmitting neuron. The transmitter in this case is partially depolarizing the second cell. If these EPSPs are of sufficient intensity or can summate, a series of action potentials may be generated in the adjacent neuron. This kind of transmission depends on temporal summation because it requires closely spaced intervals of subthreshold excitation. When two transmitting neurons synapse with a third and neither alone produces threshold stimulation, the two acting together may generate an action potential. This type of transmission is known as spatial summation. It is important to understand that most synaptic transmission probably depends on both temporal and spatial summation, since at any given synapse there are many neuronal processes involved.

Synaptic delay refers to the interval between the arrival of the action potential at the telodendria and the initiation of an action potential in the second neuron. This delay is the result of (1) the time it takes for the vesicles to discharge the transmitter substance following the arrival of an action potential, (2) the delay involved in diffusion of the transmitter across the cleft between the two neurons, and (3) the time needed to initiate the action potential on the receptor neuron.

Inhibitory postsynaptic potentials (IPSPs) may prevent threshold transmission from taking place at the synapse by hyperpolarizing the dendrites of the receptor neuron. A moment's thought will demonstrate the necessity for some kind of inhibitory process within the nervous system. If, for example, the arm is strongly flexed, the biceps muscle is contracting; but at the same time, the triceps must relax. On the other hand, if the arm is then suddenly extended, the contraction of the biceps must be inhibited immediately. This process is called reciprocal inhibition. Inhibition of muscles clearly demands inhibition at synapses, and this has been provided for by inhibitory postsynaptic potentials. These are the result of hyperpolarization of the cell, and this in turn is brought about by a change in the permeability of the cell membrane. It has been suggested that the inhibitory transmitter in this case causes the membrane to become more permeable to potassium ions ($K+$) and chlorine ions ($Cl-$). The usual influx of sodium ions ($Na+$) does not take place. The potassium ions flow out of the cell membrane and the chlorine ions inward thus causing hyperpolarization of the neuron (Fig. 2-13).

It should be emphasized that at any given synaptic site in the nervous

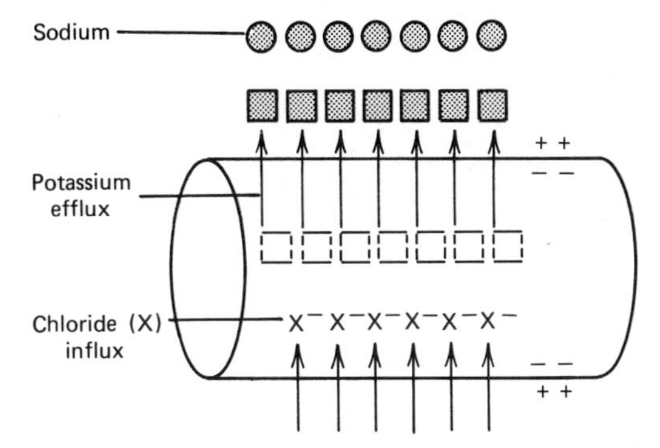

FIGURE 2-13. Hyperpolarization of the postsynaptic membrane.

system both excitatory and inhibitory processes occur together. This reflects the behavior of our muscular and glandular systems in which one-half of our activities might be characterized as "GO" and the other half as "STOP."

3

THE SPINAL CORD AS A CONDUCTION SYSTEM

TRUNK LINES TO HEADQUARTERS

During the 1972 United States presidential campaign, millions of television viewers saw one of the candidates, Governor George Wallace of Alabama, struck down in a supermarket parking lot by a would-be assassin's bullet. The candidate fell to the ground, his legs paralyzed and the lower part of his body lacking sensation. The bullet had lodged in his spinal column in the lumbar region, the shock destroying the delicate tracts of the spinal cord. From the time of the assault, Governor Wallace was confined to a wheelchair, a paraplegic, with no hope of recovery of lost spinal functions. Dramatic and tragic events such as this demonstrate the paramount importance of the spinal cord as a conduction and control center for the body.

The bundles of neurons that travel up and down the cord may be divided into three basic types: (1) motor, (2) sensory, and (3) intersegmental. Here we need only briefly note that intersegmental neurons are those that connect one level of the cord with another. Because of their importance in reflexes and voluntary motor activities, we shall consider them in detail in the chapter to follow on the integrative action of the spinal cord.

Motor pathways serve muscles and glands. We have also chosen to defer detailed discussion of the neural control of the skeletal muscle system to the following chapter and of the neural control of the glands and smooth muscles to Chapter 5, which is concerned with the autonomic nervous system. In this chapter we shall trace only descending pathways for the skeletal muscles.

Sensory pathways ascending the spinal cord serve what is called the somesthetic sense (*soma,* body; *esthesis,* feeling), or sensations of touch, pressure, temperature, and pain that arise from the skin or deeper tissues. Also included among the somesthetic senses is kinesthesis (*kinein,* to move; *esthesis,* feeling), the sense of position and movement in various parts of the body. Kinesthesis is often used synonymously with proprioception, a term derived from the fact that receptors for proprioception are defined as those within the tissues proper. These are stimulated by the activities of those tissues. Clearly, kinesthetic sensations arising as they do from muscles, tendons, and joints fit the definition; however, many authorities also include among the proprioceptors the receptors of the inner ear that mediate the sense of equilibrium (the semicircular canals, utricle, and saccule). Here we shall be concerned only with the kinesthetic receptors, deferring discussion of the sense of equilibrium to Chapter 8.

Before describing the pathways for somesthetic sensations, we will consider the types of receptors involved in the detection and transduction into nervous impulses of the stimuli for these sense modalities.

TACTILE SENSITIVITY: SKIN DEEP

The primary tactile sensations are touch, pressure, and vibration. By touch we ordinarily mean the experience of light stimulation on the surface of the skin. By pressure sensitivity we refer to activation of receptors deeper in skin and underlying tissues. Vibration is aroused by relatively rapid, repetitive stimulation. It may seem surprising to discover that neuroscientists are not yet in agreement about how many specialized receptors serve tactile sensitivity. Some upon which there is general agreement are shown in Figure 3-1.

Free nerve endings are found in the skin and many other bodily tissues. They are especially numerous in the cornea of the eye where no other type of nerve ending can be found. Many neurophysiologists believe that free nerve endings may mediate sensations of pain as well as touch.

Endings associated with hair follicles are sensitive to even the slightest displacement of hairs. These serve to make us aware of light contact as well as any movement of objects across the surface of the body.

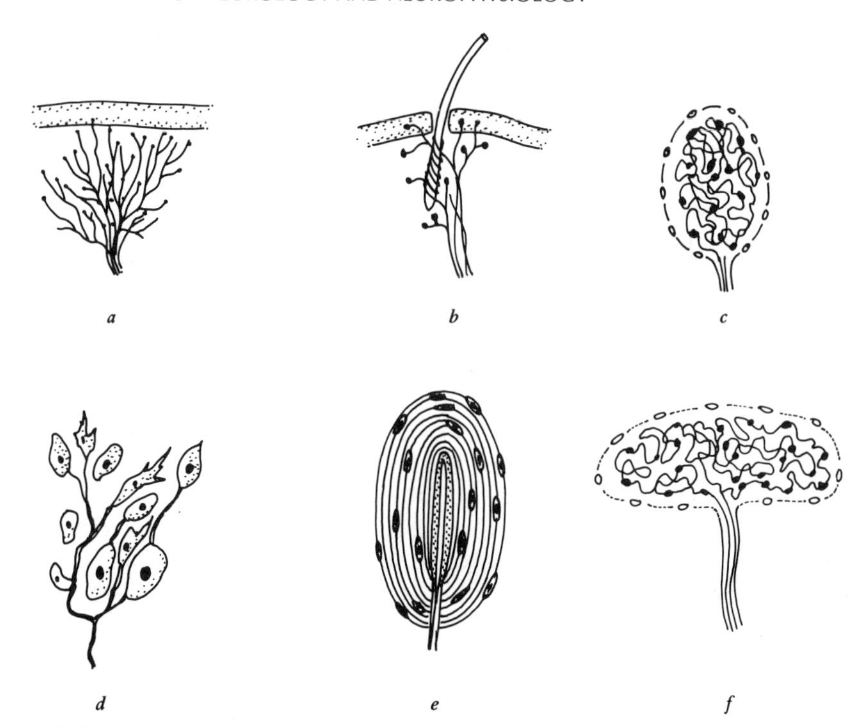

FIGURE 3-1. The principal cutaneous receptors. *a,* Free nerve ending; *b,* Tactile receptor; *c,* Meissner's corpuscle; *d,* Merkel's tactile disc; *e,* Pacinian corpuscle; *f,* Ruffini end organ.

Meissner's corpuscles (named after George Meissner, 1823–1893, a German histologist) are abundant in the nonhairy areas of the skin, particularly the finger tips, and are associated with the highly developed ability of the fingers to localize stimuli precisely and to discriminate the characteristics of objects by touch.

Merkel's tactile discs (named after Friedrich S. Merkel, 1845–1919, a German anatomist) are similar to Meissner's corpuscles in that they are most numerous in the nonhairy surfaces of the body but differ from the former in adapting less readily to stimulation of constant intensity. This suggests that they mediate our awareness of objects that remain in contact with the surface of the body for long periods of time.

Pacinian corpuscles (named after Filipo Pacini, 1812–1833, an Italian anatomist), which lie somewhat deeper under the surface of the skin than the receptors discussed thus far, are stimulated by mechanical deformation of the skin that is of sufficient intensity to reach their level. They are both extremely sensitive to pressure changes and are responsive to vibrating stimulation. These corpuscles are also found in the deep tissues of the body where they function as pressure detectors.

Finally, the Ruffini end organs (named after Angelo Ruffini, 1864–1929, an Italian anatomist), which also lie relatively deep below the surface of the skin, have been identified as pressure receptors that adapt slowly and mediate pressure sensations from heavy and continuous deformations of the skin.

TEMPERATURE SENSITIVITY: A MYSTERY

Despite an intensive search over the past century, we do not know the receptors for thermal sensitivity. Many investigators have located cold and warm spots on their skin—there are many more cold spots—and they have excised the skin and studied it microscopically, searching for specialized receptors. None has been reliably and consistently found in these Spartan experiments. Some researchers have suggested that free nerve endings may be the receptors for thermal sensitivity.

We do know that temperature sensitivity depends on skin temperature. Normally the surface of the skin is about 33C. Stimuli above 33 to about 45 feel warm; those below 33 to 13 feel cold. Stimuli beyond the range of 13–45C elicit sensations of pain. However, if the skin has been exposed to cold, as when handling snow, cold water run over the hand feels warm. The skin has adapted physiologically to cold, and the threshold for warm has been changed. We also know that sensations of burning heat can be elicited by placing warm stimuli on cold spots on the skin. Similarly, very hot stimuli under certain conditions may feel momentarily cold. These sensations are called paradoxical heat and cold. Until the receptors for warm and cold have been identified, we cannot account for paradoxical sensations.

PAIN: A NECESSARY EVIL

We are all familiar with pain, and careful research has demonstrated that we all tend to have about the same threshold for feeling pain. But, depending on our personality and cultural conditioning, we react differently to pain. We know, too, that there are different kinds of pain—the superficial pain of the skin, the deep, aching pain associated with a toothache or visceral disturbances, and the sharp, burning pain that is sometimes reported as a consequence of certain diseases or injuries.

Free nerve endings have been identified as pain receptors, but there is reason to believe that other types of receptors may also take part in the detection and transmission of pain impulses. Some neuroscientists believe that intense and potentially dangerous stimuli may be capable of generating pain reactions in any type of receptor. At the present time we are not certain.

Common experience tells us that our pain sensations do not adapt readily, though we often wish they would. However, we can be and often are distracted from perceiving pain by other stimuli, as is true of the injured football player who, giving his all for alma mater, only becomes aware of a cracked bone in his foot after the game is over.

In laboratory studies of the depression of pain impulses, tones can be fed into the ears through earphones, or stimuli can be applied to the skin arousing the tactile receptors, and either of these competing sets of signals will depress pain sensations. A similar mechanism may operate in the ancient Chinese technique of acupuncture in which impulses are generated deep in the tissues by inserting and rotating fine needles. We are not yet certain. It has been suggested that all of these competitive effects may be explained by some sort of "gating" principle. That is, when bursts of afferent stimulation are ascending the spinal cord they may take control of transmission lines and brain centers blocking the transmission of other signals. The situation may be analagous to the busy signal generated in a telephone receiver when no channels are available.

Pain may be difficult to localize, and in some disorders may be referred from one part of the body to another. A possible explanation for this phenomenon is given later in this chapter.

THE KINESTHETIC RECEPTORS: MUSCLES, SINEWS, AND BONE

The kinesthetic receptors are comprised of various specialized end organs that can be found in muscles, tendons, and joints. The more important types are shown in Figure 3-2.

The muscle receptors that are sensitive to stretch will be discussed further in the chapter to follow on reflexes and motor control.

The receptors in joints consist of the Ruffini endings and Pacinian corpuscles found in tissues around the joints and are activated by rotation of the joints. Both types of endings are extremely sensitive to even minute changes in movement or rotation.

The Golgi tendon receptors located in ligaments around the joints are sensitive to tension or torsion of these structures.

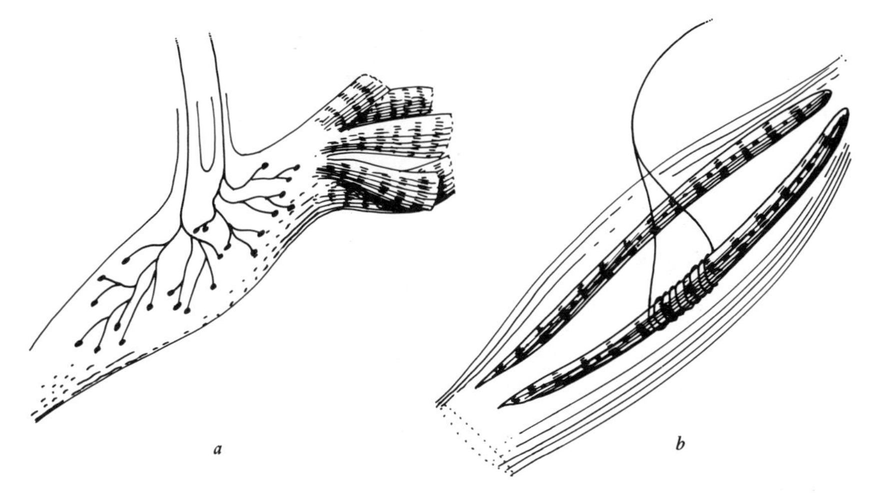

FIGURE 3-2. Two common kinesthetic receptors. *a*, Golgi tendon organ; *b*, Muscle stretch receptor.

THE VIEW FROM ABOVE: THE SPINAL COLUMN IN CROSS SECTION

To begin our study of the spinal cord as a bundle of conduction pathways, it will be helpful to look at a cross section of cord as it appears in the midthoracic region surrounded by its supportive tissues and the bony vertebral column. As Figure 3-3 shows, the spinal cord itself appears as a small, ovoid structure whose central portion is a butterfly-shaped mass of gray matter surrounding a central canal. The central portion of the cord appears gray because it is made up largely of the cell bodies of neurons which, because they contain chromatin, appear darker than the surrounding bundles of axons and dendrites that make up the white matter. This fundamental distinction between white and gray matter holds at all levels of the central nervous system, with gray matter indicating the presence of masses of cell bodies and white matter the presence of dendrites and axons that interconnect various centers.

Anatomists speak of the large bundles of neurons that make up the posterior white matter as the posterior funiculi or, if singular, funiculus (*funiculus,* little cord, or bundle). The two posterior funiculi are separated by the posterior median septum (*saepes,* fence). Similarly, the anterior funiculi make up the anterior white matter and the lateral funiculi the large areas of white matter on either side of the cord. When the anatomist wishes to speak of the "wings" of the butterfly-shaped gray matter, he uses the term "horns." The posterior horns are toward the

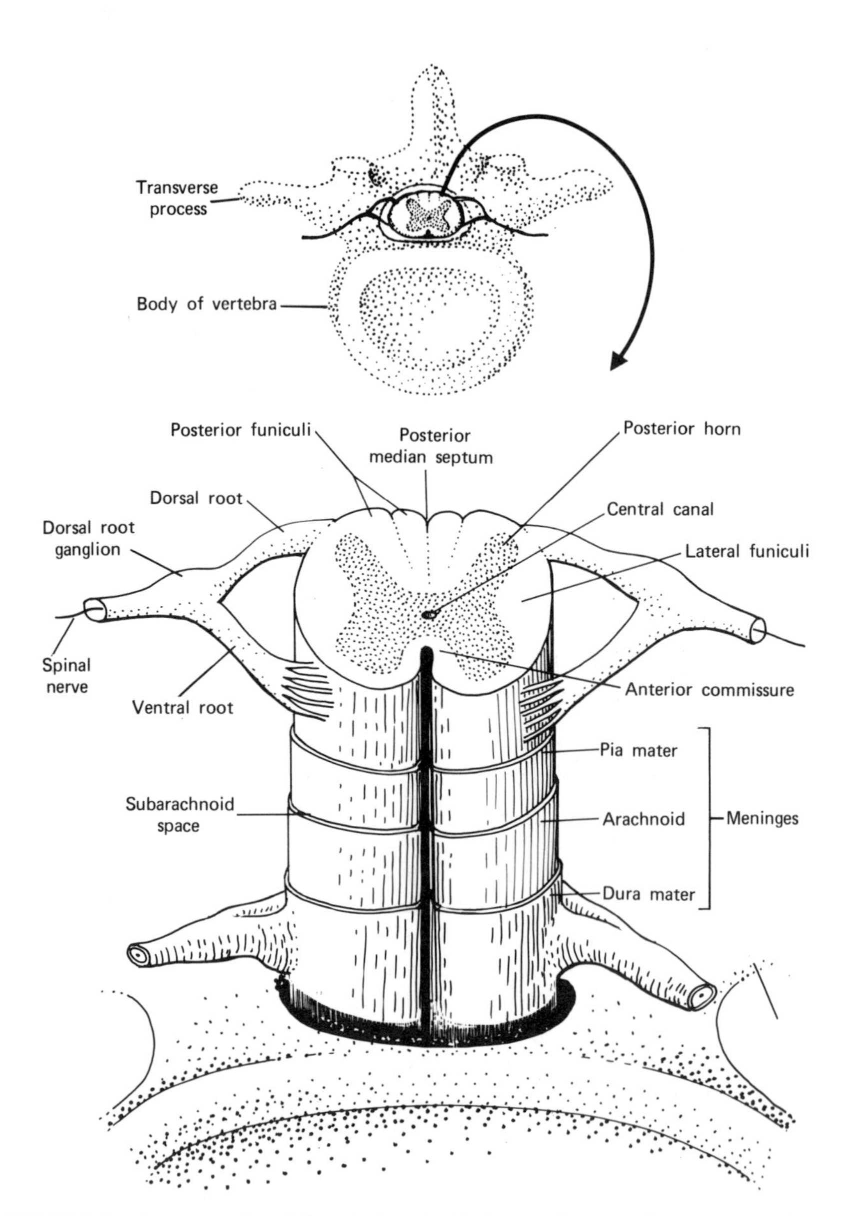

FIGURE 3-3. A cross section of the spinal cord with the spinal roots and meninges in relation to a vertebra.

back, the anterior toward the front. Finally, we should note that just anterior to the gray matter around the central canal is a white area known as the anterior commissure (*commissura,* a joining together). This is the place where connecting fibers cross from one side of the cord to another.

The outer circumference of the spinal cord is surrounded by three membranes collectively known as meninges (*mennix,* membrane). The delicate, innermost layer in intimate contact with the spinal cord is called the pia mater. The second layer is the arachnoid, so named because it is connected to the pia mater by fine, weblike filaments. The space between the pia mater and arachnoid membranes is the subarachnoid space. This space is filled with a clear fluid that is also found in the space between the skull and brain and in the ventricles or hollow chambers within the brain itself. It is called the cerebrospinal fluid and is secreted by special cells in the ventricles of the brain. One role of the cerebrospinal fluid is believed to be protective in that it keeps the delicate tissues of the central nervous system away from direct contact with the surrounding skull and vertebral column. It may also have a nutritive function, since it contains glucose, proteins, and various electrolytes. For this reason it provides an excellent medium for various pathological organisms that find their way into the central nervous system. We shall have more to say about the cerebrospinal fluid when we discuss the ventricles of the brain in subsequent chapters.

The dura mater is the tough, fibrous outermost layer of the meningeal tissues. It serves as periosteum (the smooth covering found over bones) for the inner aspect of the skull and vertebral column and also functions as a protective envelope for blood vessels, the delicate tissues of the other meninges, and the underlying nerve cells.

Figure 3-3 also shows how the spinal nerves originate as dorsal and ventral roots of the spinal cord before joining to form a common trunk leading out of the vertebral column. The dorsal roots are sensory in function and are made up of the stream of incoming sensory neurons whose cell bodies lie in the swellings just outside the spinal cord called the dorsal root ganglia (*ganglion,* swelling or enlargement). The ventral roots are motor in makeup, consisting of the axons of neurons whose cell bodies lie in the anterior gray matter of the cord and which stream outward to innervate the muscles and glands.

Spinal nerves form 31 pairs—8 cervical, 12 thoracic, 5 lumbar, 5 sacral, and 1 coccygeal (Fig. 3-4)—and are named after the corresponding vertebra from which they exit. It may be noted that the cord itself does not reach beyond the first lumbar vertebra, the remaining space being filled with the lumbar, sacral, and coccygeal nerves which, because

EXPERIMENT III

BELL AND MAGENDIE PROVE THE LAW OF SPINAL ROOTS

The law of the anatomical and functional discreteness of the motor and sensory nerves as they enter the cord is known as the Bell-Magendie law for its discoverers, Sir Charles Bell (1774–1842), a brilliant British physiologist and surgeon, and François Magendie (1783–1855), a famous French physiologist. Working independently, these two investigators proved that the dorsal or posterior roots entering the cord are sensory in function and the ventral or anterior roots are motor in function.

Using animals as subjects, the cord was sectioned and the posterior roots stimulated at *(1a)* and *(1b)* as shown in the illustration. Stimulation at *(1a)* resulted in a reflex motor response in the muscles on the periphery. Stimulation at *(1b)* gave no response. In a second animal the cord was sectioned and stimulation applied at *(2c)* and *(2d)* as shown in the illustration. A motor response was obtained at *(2d)* but not at *(2c)*.

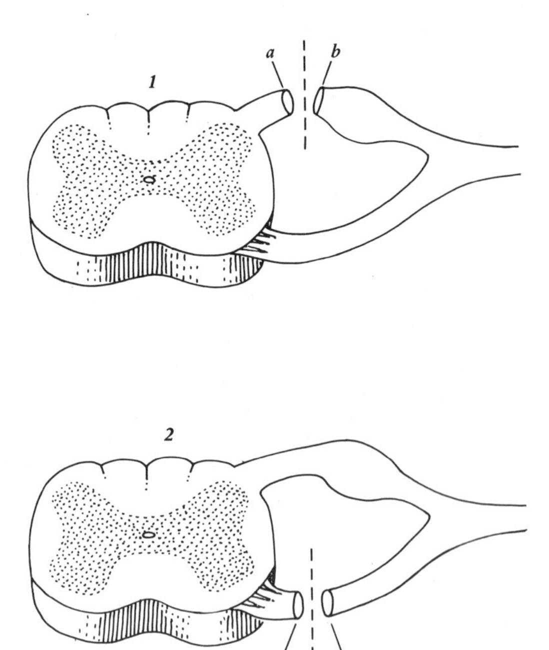

In the first experiment an impulse was generated which traveled into the cord over the posterior root neurons that synapse with motor neurons leading to the voluntary muscles. Upon arrival of the impulse, the voluntary muscles responded. Stimulation at *(1b)* had no effect, since the impulses so generated could only travel backward to receptor sites where they would die out.

In the second experiment, stimulation at *(2c)* resulted in no response, since it could only generate impulses that traveled backward to the cord to die out at the synapse. Stimulation at *(2d)* generated nervous impulses that traveled directly to the muscles where they excited contractions.

Sir Charles Bell was also noted for his research on kinesthesis or the muscle sense, and for describing the fact of reciprocal innervation of muscles. Magendie was noted for research on blood as well as for his independent discovery of the law of spinal roots.

of their bushy-tailed appearance as they course downward, were given the colorful name, cauda equina (*cauda,* tail; *equus,* horse). It may also be observed that the spinal nerves form three large networks known as the brachial, lumbar, and sacral plexuses (*plexus,* to turn or twist). These networks supply the arm, shoulders, and the lower region of the body—hips, pelvis, and legs.

SLICES AND PATCHES: DERMATOMES

Even a superficial study of the vertebral column and its associated spinal nerves demonstrates the segmental nature of bodily organization. This segmentation is, of course, not limited to man—in fact it may be more easily observed in fishes and reptiles. There is segmentation not only of the bony structures but of the distribution of the spinal nerves as well. As Figure 3-5 shows, each spinal nerve has its particular segment of the body for which it is responsible. These segments are called dermatomes (*derma,* skin; *tome,* slice or segment). The dermatomes are not so sharply divided as the illustration implies, for there is considerable overlapping among them. Moreover, the deep tissues and visceral organs are also segmentally organized in terms of their innervation by spinal nerves. Dermatomes are of importance in helping the neurologist differentiate true organic diseases of the nervous system from functional or hysterical disorders and in accounting for referred pain.

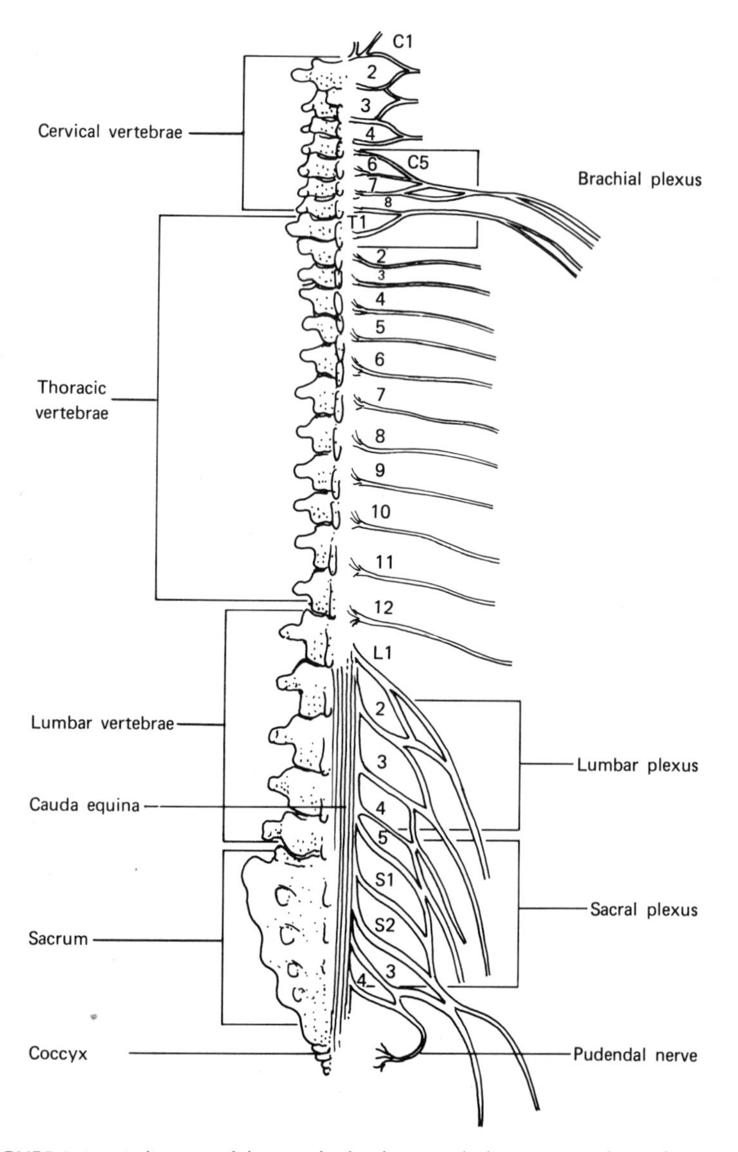

FIGURE 3-4. A diagram of the vertebral column with the 31 pairs of spinal nerves.

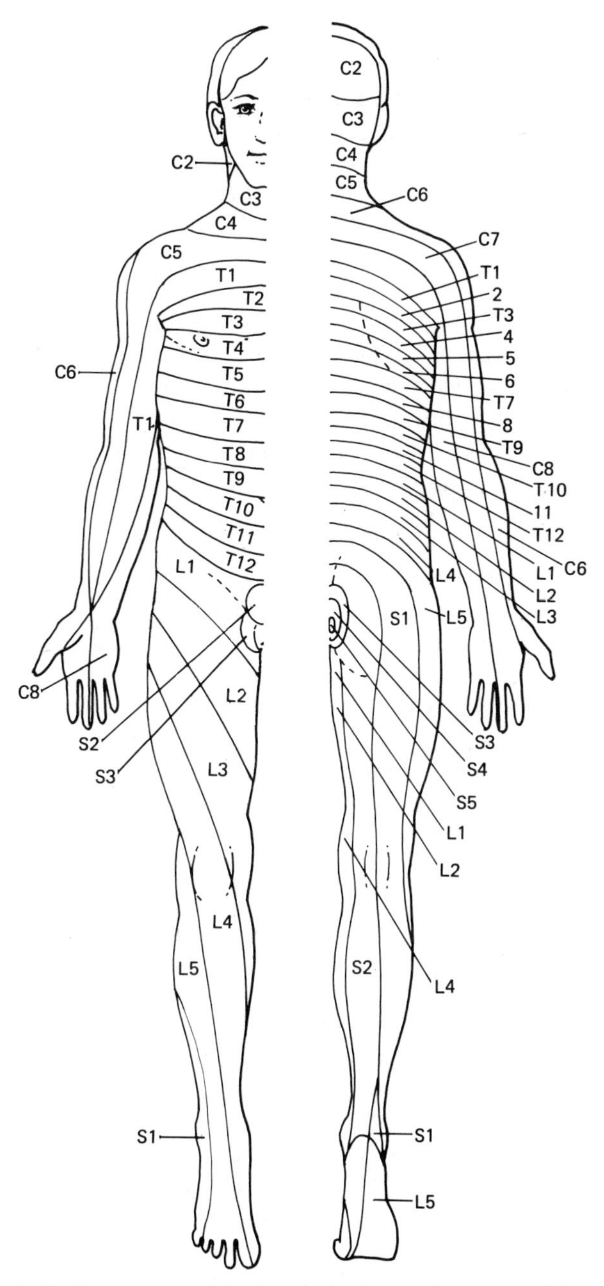

FIGURE 3-5. Dermatomes of the human body, anterior and posterior views.

Persons suffering from hysterical disorders, for example, often complain of partial or total paralysis of the limbs or of anesthesias or loss of sensation in various parts of the body. In many cases of sensory loss, they may tell the examiner that the lower part of the arm from the elbow down (or the leg from the knee down) is anesthetic and that they have no sensations in the affected part. Technically, these are called stocking and glove anesthesias, since they commonly involve areas covered by womens' formal gloves or stockings. It should be emphasized that these individuals are truly anesthetic to pin pricks and to thermal and other sensory stimuli used as tests. They are not faking or malingering. However, their problems are psychological, not neurological. As a study of Figure 3-5 shows, innervation of the arm and leg is not neatly divided at the elbow or knee but is irregularly distributed in patches.

Referred pain is as yet poorly understood but may be described as pain that is displaced away from its true source to some other location. For example, cardiac patients often complain of pain along the arm or in the neck. Sometimes pain in the diaphragm may be felt over the shoulder. Similarly, visceral pains are often poorly localized. It is possible that in some cases of referred pain the same dermatomes serve the skin and underlying tissues of the diseased organ, which is the true source of pain. It has been suggested that the mass of nervous impulses coming in from the affected organ may spill over in the spinal cord to trigger reactions in adjacent pools of neurons whose origin is in the skin. As these centrally generated impulses reach the brain, the individual misinterprets his discomfort as originating on the periphery. We are not sure that this is the mechanism of referred pain, but it is a plausible hypothesis. In any case, the clinician must be alert to the possibility that the true origin of pain may be different from the reported origin.

CUTTING THE TRUNK LINES: THE SPINAL PATHWAYS IN CROSS SECTION

Returning to our cross-sectional view of the spinal cord, Figure 3-6 shows in detail the various spinal pathways as they travel up and down the cord. These tracts do not remain as distinct as the diagram suggests. However, because we are emphasizing functional rather than anatomical distinctions, the pathways are shown as if they remained anatomically separate. It will also be noted that the chief ascending or sensory tracts are shown only on the right one-half of the figure and the descending tracts on the left. This has been done for the sake of clarity, even though both sides of the cord are, of course, identical.

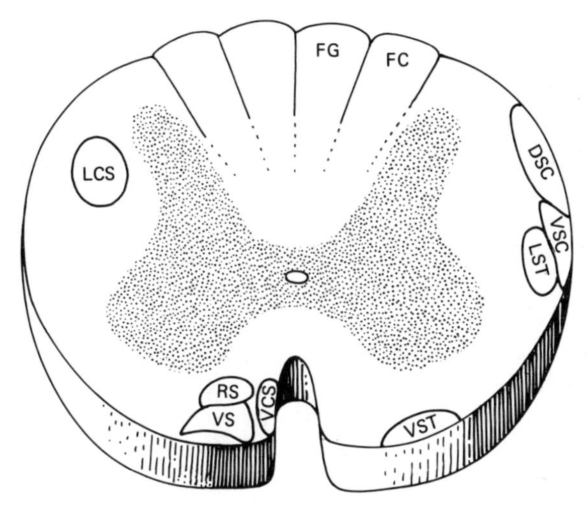

FIGURE 3-6. Cross section of the spinal cord in the midthoracic region showing the principal tracts. Ascending tracts are shown on the right, descending on the left. *Descending tracts:* LCS, Lateral corticospinal; VCS, Ventral corticospinal; RS, Reticulospinal; VS, Vestibulospinal. *Ascending tracts:* FG, Fasciculus gracilis; FC, Fasciculus cuneatus; DSC, Dorsal spinocerebellar; VSC, Ventral spinocerebellar; LST, Lateral spinothalamic; VST, Ventral spinothalamic.

Because we shall be tracing the conduction pathways in detail in the section to follow, our task here will be limited to becoming familiar with their names and general functions.

Fasciculus gracilis (*facis,* bundle; *gracilis,* slender) and fasciculus cuneatus (*cuneus,* wedge-shaped) are collectively known as the dorsal columns. Fibers in these tracts mediate touch, particularly that aspect of tactile sensitivity associated with the inability to localize stimuli and perceive twoness or astereognosis (*a-,* privation; *stereos,* solid; *gnosis,* knowledge) which is tested by determining whether the blindfolded subject can detect as separate two closely spaced points simultaneously placed on the skin. The dorsal columns also mediate conscious kinesthetic signals from joints.

Fasciculus gracilis contains fibers from the lower limbs, and it is therefore a long, slender pathway extending all the way up the cord. Fasciculus cuneatus, on the other hand, begins to appear only in the upper part of the cord, since its fibers originate in the upper trunk and neck region.

The dorsal and ventral spinocerebellar tracts carry conscious kinesthetic impulses primarily from muscle and tendon receptors to the cerebellum. We shall discuss in more detail in another chapter the control of motor activity as a joint function of the cerebrum and cerebellum working in cooperation. Because of their rich interconnections, both parts of the brain need information from ascending sensory pathways in the spinal cord about the tone and movements of the muscles, tendons, and joints.

The lateral spinothalamic tract consists of two closely associated and anatomically indistinguishable bundles of fibers that conduct impulses mediating the senses of pain and temperature.

Among the ascending pathways in the anterior part of the cord is the ventral spinothalamic tract, which conducts impulses mediating simple tactile sensitivity.

Two important descending pathways are the lateral and ventral corticospinal tracts, which conduct impulses from the motor centers in the cerebral cortex to the spinal neurons that make up the ventral roots leaving the cord for the control of muscular activity. As we shall see in more detail in the section to follow, the larger and by far more important of the two pathways is the lateral corticospinal tract, a crossed path with fibers descending on the left side of the cord having originated on the right side of the brain and vice versa. The ventral corticospinal pathway remains uncrossed until it reaches the level of the cord where it terminates. There the neurons cross to the opposite side where they synapse either directly or indirectly through internuncial neurons with the lower motor neurons of the ventral roots.

The intersegmental bundles are pathways that originate in one level of the spinal cord and terminate in another. They provide for intersegmental reflex activity such as we observe when people are walking, alternately swinging the opposite arm to the leg in motion. This reflex is a carry-over from the days when our ancestors walked on all fours.

The vestibulospinal pathways conduct impulses from centers in the midbrain, which in turn are connected to organs of the inner ear (semicircular canals and vestibule) that monitor balance and changes in bodily orientation. The contribution of these pathways is believed to be important in maintaining muscle tone and in facilitating reflex action, particularly reflex action in the extensor muscles that support the limbs and body.

The reticulospinal tract originates in the medulla in that part known as the reticular formation (*reticulum,* little net), an important network of fibers concerned with selectively relaying the streams of impulses that are constantly coming up from the lower body and down from the cortex. We shall discuss this important center more fully in Chapter 11.

Here we need only note that research shows that the reticular formation has important functions in alerting the cerebral cortex and in exerting a facilitatory effect on the voluntary extensor muscles.

THE VIEW FROM BELOW UPWARD: THE AFFERENT SPINAL TRACTS

In this section we will trace the neuronal chains that make up the principal pathways leading from the spinal cord to the brain. It will be convenient to label the neurons making up the chains with Roman numerals I, II, etc. to indicate their order. All of the sensory or afferent (*ad,* to; *ferre,* to carry) neurons making up ascending tracts have their origins in receptors either in the skin or deep within the muscles, tendons, and joints. Because these systems mediate impulses from a number of different sense modalities from all over the body rather than from specialized areas as is the case of the head receptors, they are called general senses and their corresponding pathways general afferent pathways.

As we noted in describing the spinal cord from a cross-sectional point of view, kinesthetic impulses arise from specialized receptors in the muscles, tendons, and joints. Impulses from these receptors are essential for our awareness of both the positions of the limbs in space and of movements at joints. Some of the impulses mediating kinesthesis are directed at the cerebral cortex where they give rise to conscious sensations. However, many terminate in the cerebellum and other centers where they do not result in conscious awareness of muscle, tendon, and joint sensations but instead provide the neural basis for automatic adjustments in these systems.

SOMESTHESIS TO THE CORTEX BY WAY OF FASCICULUS GRACILIS AND FASCICULUS CUNEATUS

Neuron I of these pathways originates in tactile and kinesthetic receptors. It enters the spinal cord by way of the dorsal roots (Fig. 3-7). Its cell body is located in the dorsal root ganglion; its axon enters the cord to travel up the same side by way of the tract of gracilis or of cuneatus, terminating in the nucleus of gracilis and of cuneatus in the lower medulla.

Neuron II originates in the nucleus of gracilis or of cuneatus, crosses

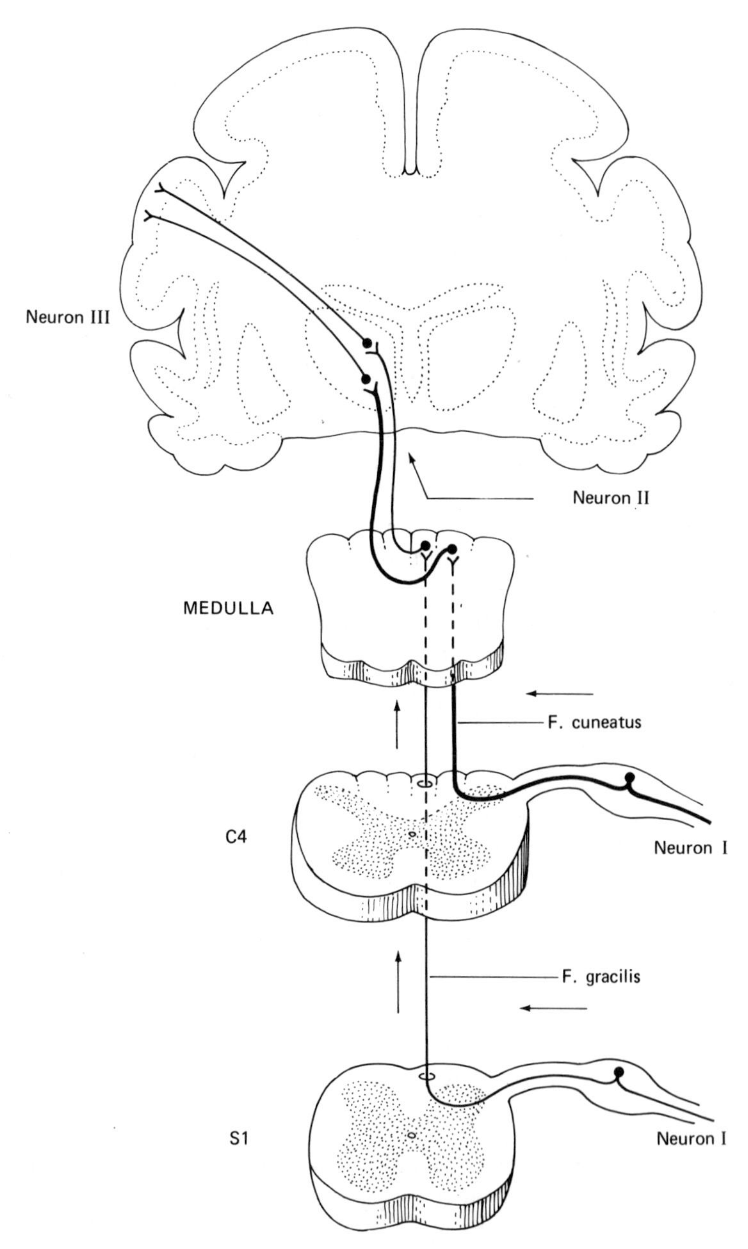

FIGURE 3-7. The tracts of fasciculus gracilis and fasciculus cuneatus.

over to the opposite side of the medulla and travels up the medial lemniscus (*lemniskos,* a strip), a narrow band of nerve fibers that ascend the medulla and terminate in the lateral thalamus.

Neuron III originates in the lateral thalamus and terminates in the somesthetic area of the cerebral cortex.

KINESTHESIS TO THE CEREBELLUM BY WAY OF THE VENTRAL SPINOCEREBELLAR TRACT

Neuron I originates from kinesthetic receptors in the muscles and tendons to enter the spinal cord by way of the dorsal root and end in the posterior horn (Fig. 3-8).

Neuron II originates in the posterior horn and ascends on either the same or opposite side to terminate in the cortex of the cerebellum.

KINESTHESIS TO THE CEREBELLUM BY WAY OF THE DORSAL SPINOCEREBELLAR TRACT

Neuron I originates in kinesthetic receptors in the muscles and tendons, and terminates in the posterior horn of the spinal cord (Fig. 3-9).

Neuron II originates in the posterior horn and ascends on the same side by way of the dorsal spinocerebellar tract to the cerebellar cortex.

TOUCH BY WAY OF THE VENTRAL SPINOTHALAMIC TRACT

Neuron I originates in sensory receptors for tactile sensitivity in the skin and enters the cord by way of the dorsal roots and ends in the posterior horn on the same side (Fig. 3-10).

Neuron II originates in the posterior horn, crosses over to the opposite side, and ascends by way of the ventral spinothalamic tract to the thalamus.

Neuron III originates in the thalamus and terminates in the somesthetic region of the cerebral cortex.

Finally, among the ascending tracts we shall trace the combined paths for pain and temperature by way of the lateral spinothalamic tract.

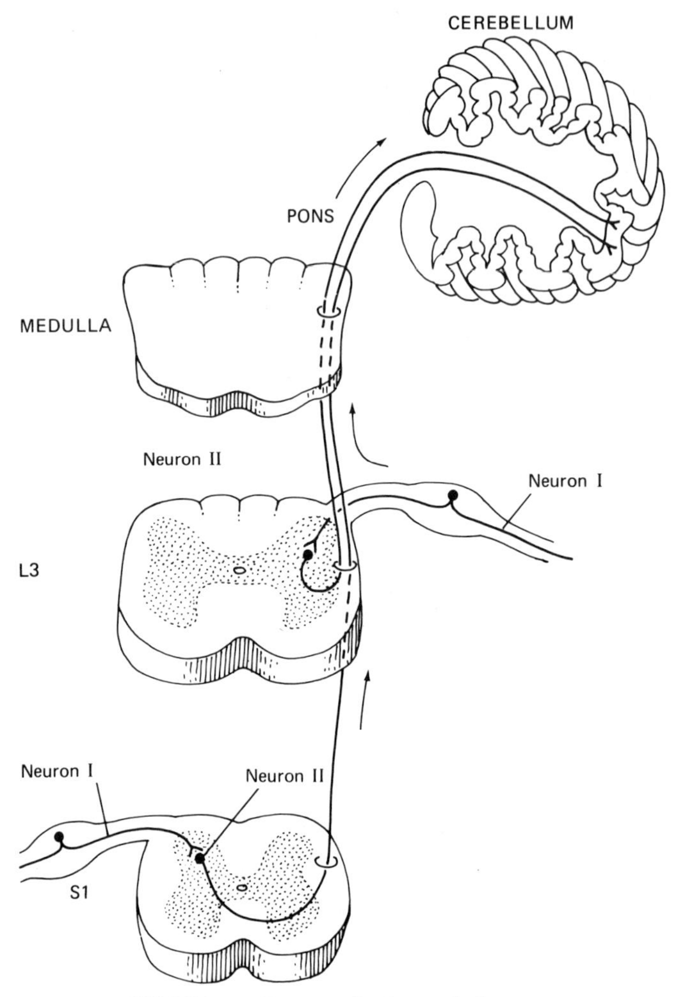

FIGURE 3-8. The ventral spinocerebellar tract.

Neuron I (Fig. 3-11) originates in receptors in the skin (or in the case of pain in deeper tissues), enters the spinal cord by way of the dorsal root, and ends in the posterior horn on the same side.

Neuron II originates in the posterior hron, immediately crosses the cord, and ascends by way of the lateral spinothalamic tract to the lateral nucleus of the thalamus.

Neuron III originates in the lateral thalamic nuclei to terminate in the somesthetic area of the cerebral cortex.

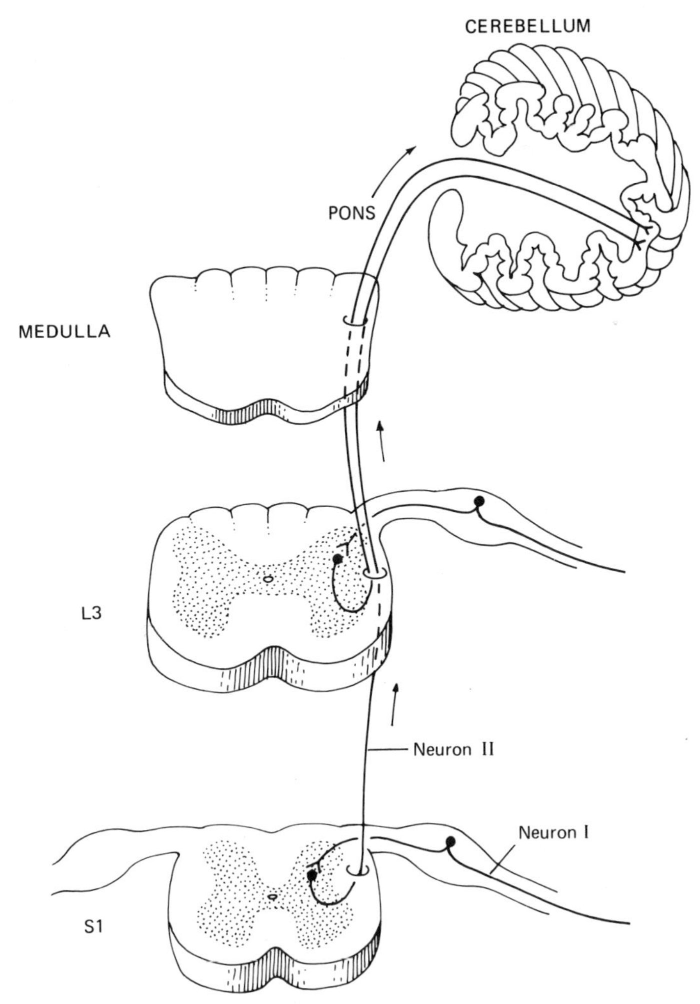

FIGURE 3-9. The dorsal spinocerebellar tract.

THE VIEW FROM THE TOP DOWNWARD: THE DESCENDING SPINAL TRACTS

There are two major pathways that originate in the motor area of the cerebral cortex and travel down the spinal cord to end in the anterior horn cells, where they synapse with neurons that travel out over the ventral roots to innervate muscles. The neurons that conduct impulses

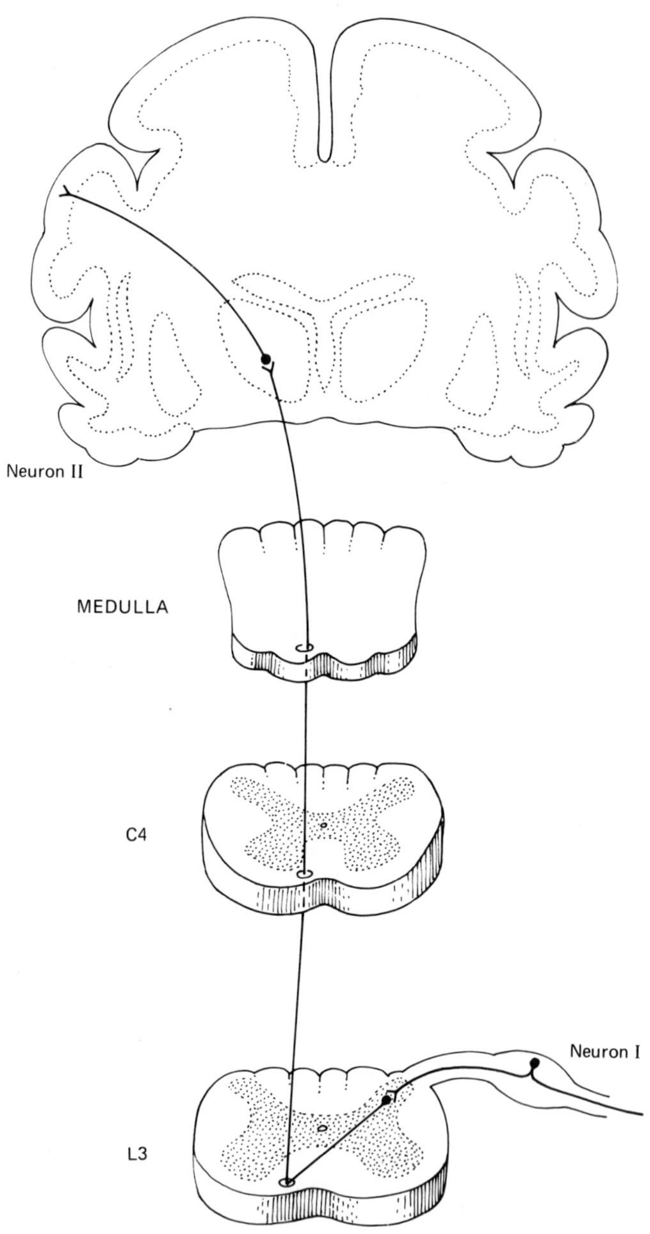

FIGURE 3-10. The ventral spinothalamic tract.

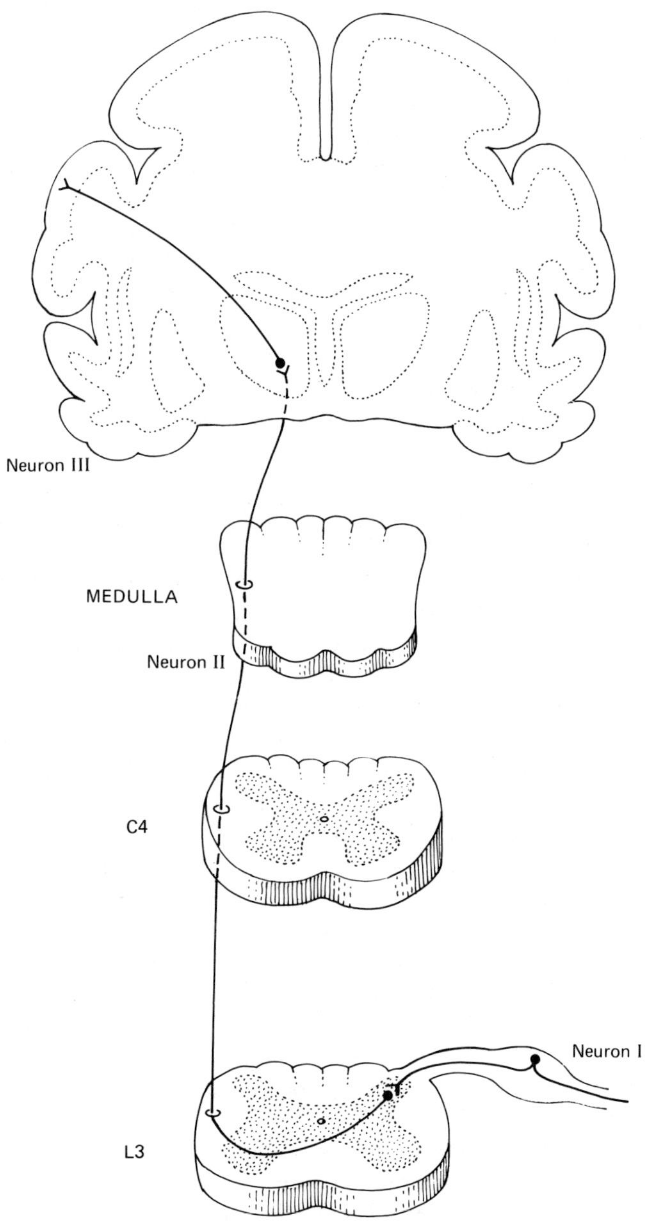

FIGURE 3-11. The lateral spinothalamic tract.

CASE III

CORDOTOMY FOR THE RELIEF OF INTOLERABLE PAIN

Mrs. C.A., a forty-year-old woman, was admitted to the medical center hospital because of intractable pain in her right leg. Three years previously she had undergone a radical operation for removal of a carcinoma of the rectum necessitating a colostomy which functioned satisfactorily. However, about a year and one-half after the operation, the patient began to suffer severe pain in the right hip. Radiological examination revealed that the malignancy had spread to her pelvis and right femur. Cobalt treatments and chemotherapy were initiated to slow further metastasis, or spread of the malignancy, and an attempt was made to control the pain with aspirin, percodan, and occasional injections of morphine. In spite of increasing dosages and frequency of administration of these drugs, they became ineffective in controlling pain. Because of the severity of her pain, the patient avoided moving her right leg, and the muscles had begun to atrophy. Moreover, there was poor coordination of the knee, ankle, and toes of the right leg as a result of restriction of voluntary movement. Radiological examination also revealed degenerative changes of the right femur and pelvis because of the carcinoma.

On the basis of these findings, it was decided to interrupt the spinal tracts in the left anterolateral quadrant of the cord in the region of the fourth thoracic vertebra, thus cutting the lateral spinothalamic tract which mediates pain. Accordingly, the spinal cord was exposed with the patient in a prone position. The fourth dorsal sensory root was cut to allow rotation of the cord so that the anterior portion became accessible. A cordotomy blade was inserted, and a major portion of the left anterolateral quadrant of the cord was severed.

Following recovery from surgery, pain was completely absent in the right leg. Superficial pain (to pin pricks) was lost on the right side below the level of the fifth thoracic segment, but there was no loss of motor function. Although the patient's prognosis was poor because of the metatasizing carcinoma, she had been made comfortable, and her motor function restored.

In cases involving more restricted carcinomas that are giving rise to intractable pain, an operation known as rhizotomy (*rhizos,* root; *otomy,* cut) is performed in which a number of sensory roots leading into the spinal cord are cut in order to block pain impulses from entering.

from the cerebral cortex to the spinal cord are known as upper motor neurons and those that conduct impulses from the spinal cord to the muscles as lower motor neurons. Collectively this system is called the corticospinal or pyramidal system. The term pyramidal refers to the fact that the relatively large motor neurons of the cerebral cortex travel from the brain to the spinal cord over pyramidal-shaped prominences in the medulla (see Chap. 9). As we shall discover when we discuss the disorders of the spinal cord, it makes a great deal of difference which of these two types of neurons, upper or lower, is affected.

THE LATERAL CORTICOSPINAL TRACT

About 80 percent of the upper motor neurons travel over this path. Consequently, it is the most important of the two motor pathways from cortex to cord.

Neuron I originates in the motor area of the cerebral cortex and descends to the region of the medulla where most of the fibers cross at a place called the decussation (*decussatus,* in the form of an X) of the pyramids (Fig. 3-12). There the intertwining fibers of the tract cause a noticeable swelling. Neuron I continues down the lateral corticospinal tract to end in the posterior horn of the spinal cord where most fibers synapse either directly or through internuncial neurons with the lower motor neurons. Some may synapse with fibers from the posterior funiculi, apparently as a means of providing feedback to the sensory system.

Neuron II originates in the anterior horn of the spinal cord to terminate in the muscle endplates.

THE VENTRAL CORTICOSPINAL TRACT

Neuron I originates in the pyramidal cells of the motor region of the cerebral cortex and descends the spinal cord by way of the ventral corticospinal tract to cross to the opposite side before synapsing either directly or through internuncial neurons with the lower motor neurons (Fig. 3-13). In this way completely contralateral control of the muscle system is assured, since the crossing of the lateral corticospinal tract in the medulla necessitates the eventual crossing of the ventral corticospinal tract before synapsing with the lower motor neurons. It is not known why motor control of the muscle system is contralateral.

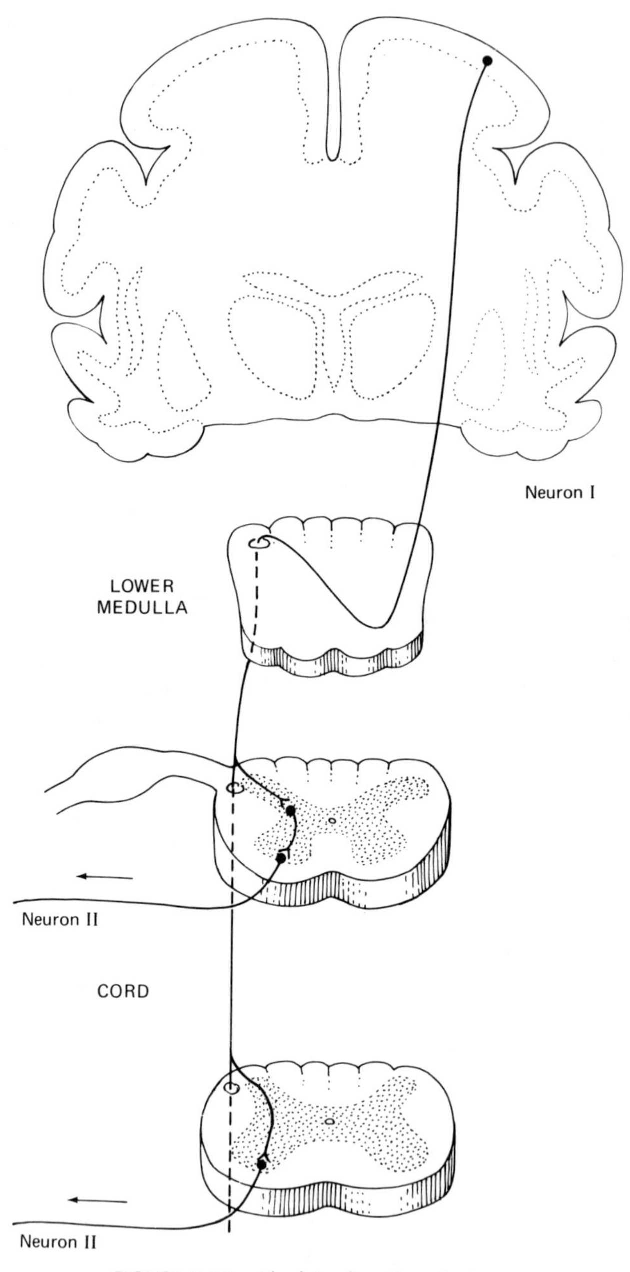

FIGURE 3-12. The lateral corticospinal tract.

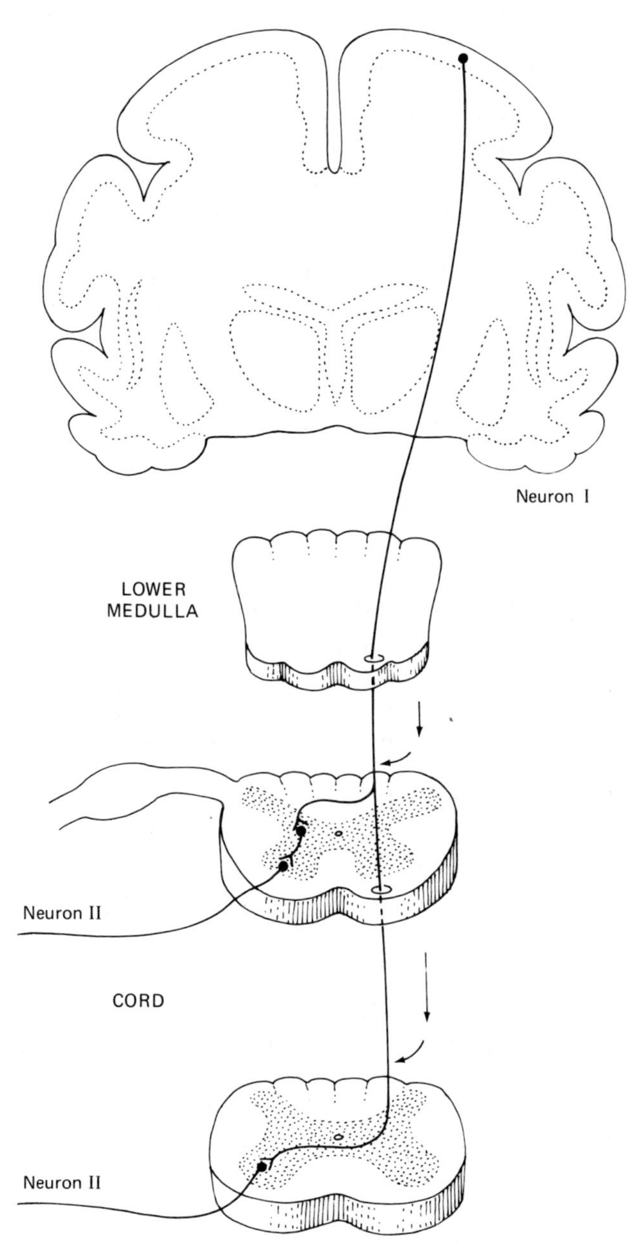

FIGURE 3-13. The ventral corticospinal tract.

OTHER TRACTS ASSOCIATED WITH MUSCULAR CONTROL

There are a number of pathways that influnce the motor system besides the pyramidal tracts, which function to provide direct cortical control over the voluntary muscle system. Structures other than the pyramidal tracts that exert influence over the motor neurons are collectively called the extrapyramidal system. Fibers composing this system arise from a variety of centers in the brain including the cerebral cortex anterior to the primary motor region, the vestibular nuclei in the medulla, which connect, in turn, with the organs for balance in the inner ear and cerebellum. As typical of pathways of the extrapyramidal system, we will consider briefly the reticulospinal and vestibulospinal pathways.

THE RETICULOSPINAL TRACT

The reticular formation is a network of fibers in the lower part of the brain that gives rise to a tract that exerts a facilitating influence on the extensor muscles of the legs and flexor muscles of the arms.

Neuron I originates in the reticular formation and travels down the reticulospinal tract on the same side to terminate in the anterior horn of the spinal cord where it synapses directly or through an internuncial neuron with the lower motor neurons of the ventral root (Fig. 3-14).

The reticulospinal system also has inhibiting components that are routed down the spinal cord by way of the lateral corticospinal tract to synapse with lower motor neurons, which form the ventral roots. This may occur either directly or indirectly through internuncial neurons.

THE LATERAL VESTIBULOSPINAL TRACT

Neuron I originates in the lateral vestibular nucleus and descends on the same side by way of the lateral vestibulospinal tract to end in the anterior horn. There it synapses either directly or indirectly with lower motor neurons (Fig. 3-15). The function of the lateral vestibulospinal tract is to reinforce muscle tone in reflex activity.

THE MEDIAL VESTIBULOSPINAL TRACT

Neuron I originates in the medial vestibular nucleus in the brainstem and crosses over to the opposite side to end in the anterior horn. There it

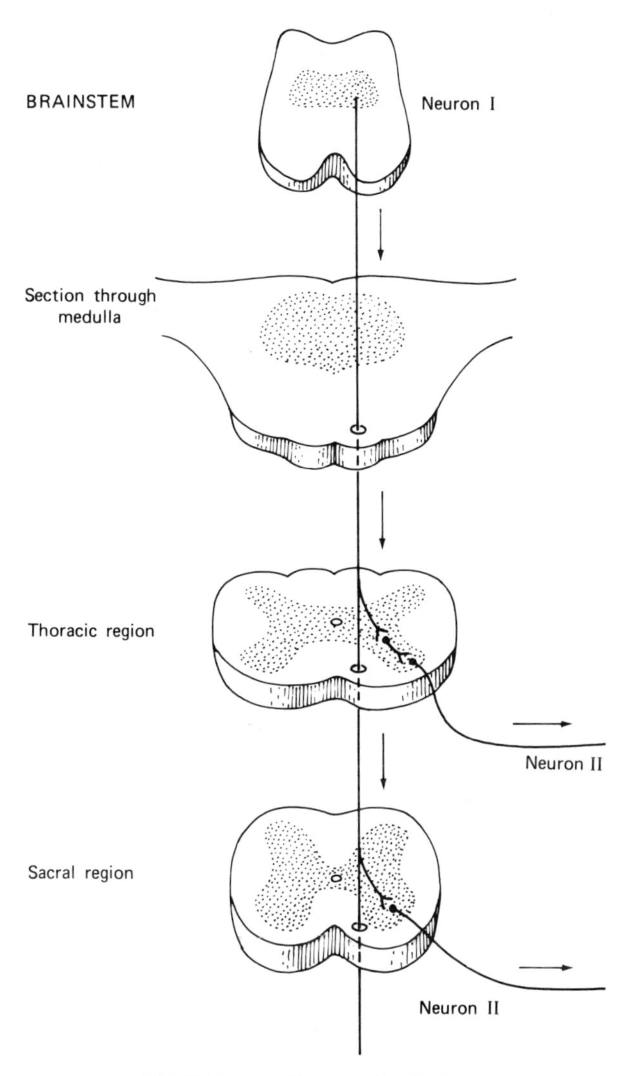

FIGURE 3-14. The reticulospinal tract.

synapses either directly or indirectly with lower motor neurons (Fig. 3-16). This tract does not descend below the cervical level in man and is believed to be concerned with the extrapyramidal control of the muscles of the neck and head.

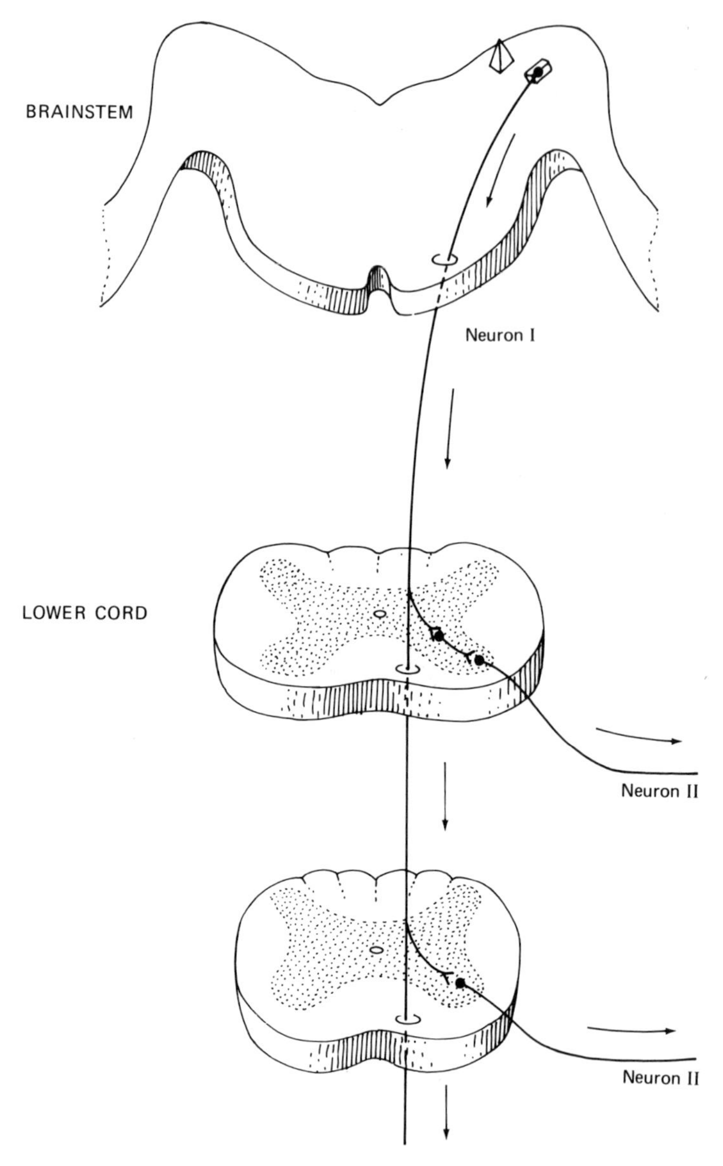

FIGURE 3-15. The lateral vestibulospinal tract.

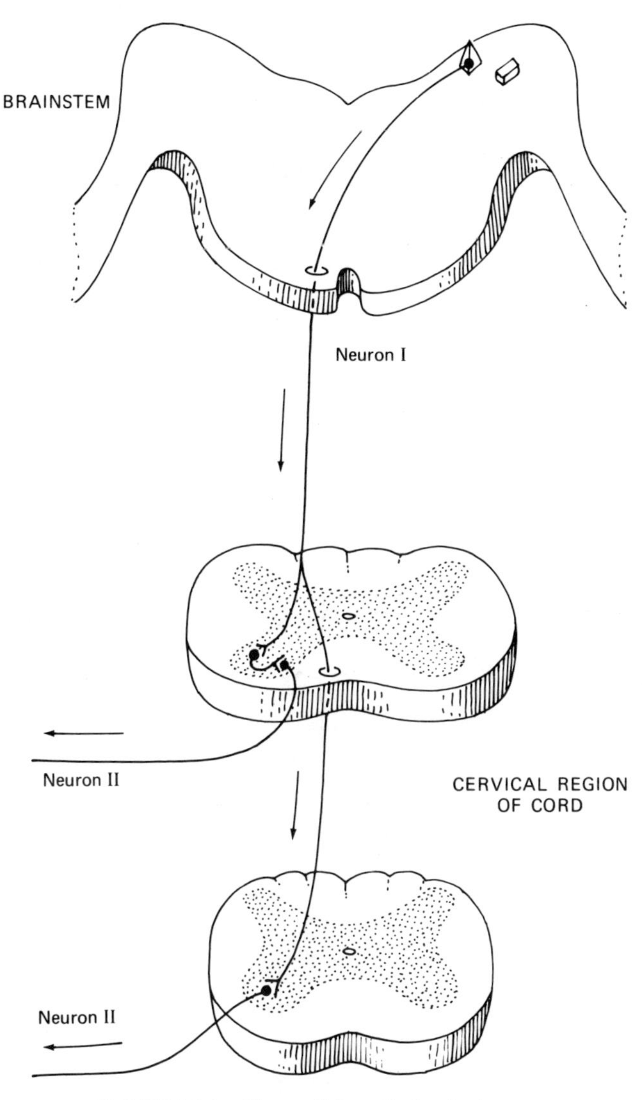

FIGURE 3-16. The medial vestibulospinal tract.

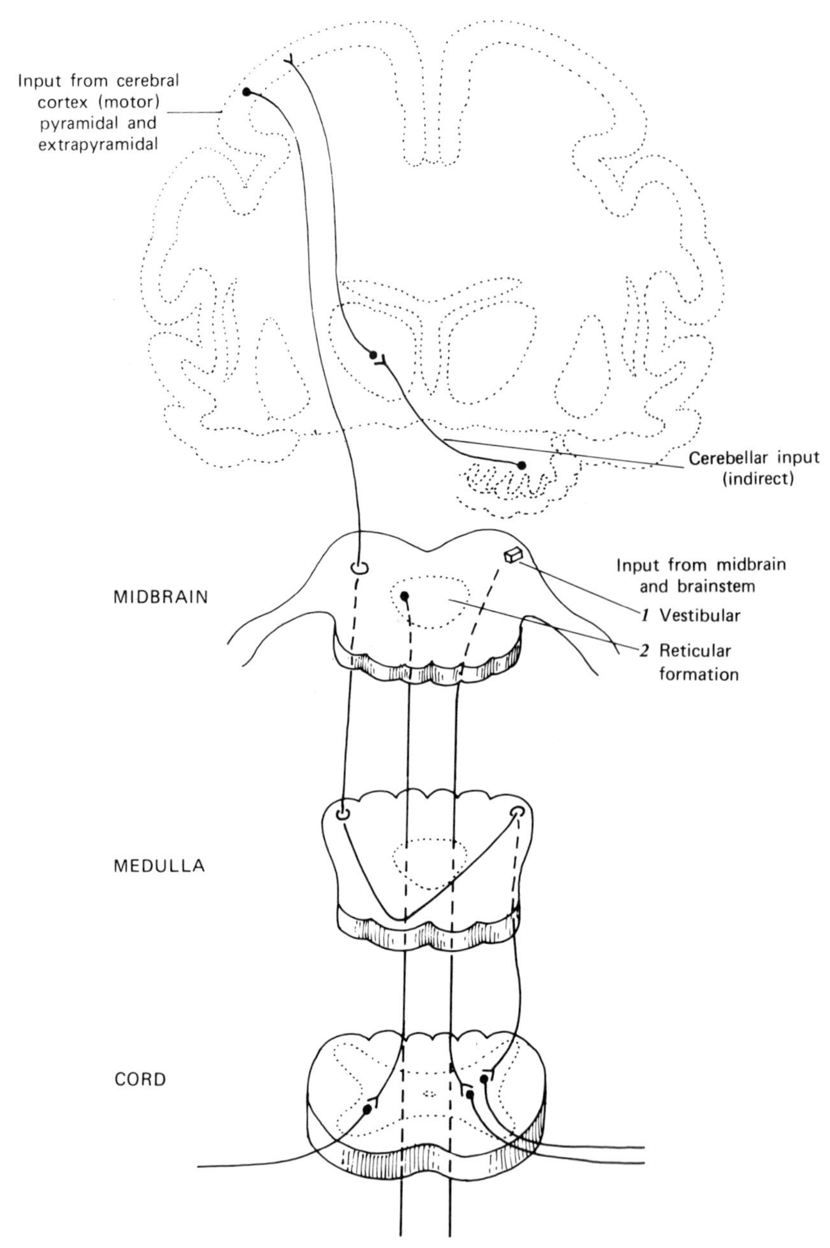

FIGURE 3-17. The final common path. A schematic representation of the contribution from the motor cortex, extrapyramidal system, cerebellum, vestibular system, and reticular formation of the brain stem to the lower motor neurons.

WHERE IT ALL ENDS: THE FINAL COMMON PATH

We shall have more to say about the control of muscular activity in Chapter 4, but it seems appropriate to end this section by introducing the concept of the final common path. This concept refers to the fact that the lower motor neurons of the anterior gray matter, which leave the cord by way of the ventral roots, are a common path for a number of influences originating in a variety of centers. Figure 3-17 shows how contributions from the pyramidal centers in the cerebral cortex join with those from the extrapyramidal centers of the cortex, from the brainstem, and from the cerebellum. To illustrate how this system functions, we may take the everyday example of an individual stooping over to pick up an object. The action is not a simple one of using the extensor muscles to reach for the object and the flexors to grasp and retrieve it; rather it involves the inhibition of the flexors during reaching, the compensatory extensor contractions of the leg muscles while reaching and leaning forward to avoid falling, the contribution of the vestibular system to the maintenance of balance, the influence of the cerebellum on postural adjustments, and the inhibition of the extensors of the arm while grasping and retrieving the object. All of these influences are finally summated on the motor neurons as they leave the cord over the anterior roots.

4

THE INTEGRATIVE ACTION OF THE SPINAL CORD

WITHOUT THINKING

There is an old story about a centipede—a species whose locomotor skill is legendary—which, when asked how it managed to keep track of a hundred legs all going simultaneously, became thoughtful about the operation of its appendages and "fell helpless in the ditch, unable now to run." Like the centipede most of us are not conscious of our locomotor behavior, at least under ordinary conditions. We do, of course, "order" ourselves to walk or run, but once underway much of the complex business of flexing and extending the muscles runs off more or less automatically. In fact, only those who have the requisite knowledge of anatomy and neurophysiology can appreciate how very complex a process locomotion actually is and how difficult it would be to control it consciously.

Consider also the case in which someone accidentally places a hand on a hot surface. Quickly, he or she removes it with a sharp flexion of the arm. Common sense says that the hand is removed because pain is felt. Precisely timed studies of the speed with which the action is carried out, however, show that the hand is removed *before* the pain can be felt consciously. This kind of response is called a reflex (*re,* back; *flectere,* to bend). This and other reflexes are stereotyped or fixed motor responses to sensory stimulation. They vary little from time to time, or even from individual to individual. Many reflexes are protective in nature—blinking to a sudden, unexpected object thrust toward the eyes, coughing, vomiting, tearing in response to a cinder on the cornea, withdrawal of a limb to a painful stimulus—all are in the service of the individual as a protection against noxious or dangerous stimuli. Other reflexes, as we shall find, play a significant part in locomotor behavior, postural adjustments, and to a limited degree as a background to voluntary behavior. This widespread utilization of reflexes frees the brain from the necessity of detailed conscious guidance of the muscle systems involved. Finally, we should not forget that reflex glandular secretions play a major role in sweating, in digestion, in cardiac and respiratory regulation, and in sexual responses. We shall consider some of these visceral reflexes in more detail when we describe the functions of the autonomic nervous system in Chapter 5.

Because of the importance of reflexes in understanding a variety of behavior patterns, and since they are useful diagnostic signs of the condition of neural pathways, neurophysiologists have done extensive research on the mechanisms of motor reflexes. In doing so they have made considerable progress in understanding the integrative action of the spinal cord.

A SWIFT KICK: THE KNEE JERK, A MYOTATIC REFLEX

The simplest type of integrative action in the spinal cord is the monosynaptic reflex, such as may be found in the typical myotatic (*mys,* muscle; *tais,* stretching) reflex that can be observed in the knee jerk, a reflex made famous in the neurologist's examining room. The examiner strikes the subject's dangling leg just below the patella or knee cap. This action causes a sudden stretching of the muscle of the thigh (quadriceps) because the tendon that attaches it to the lower leg has been stretched by the blow. The response is a quick contraction of the muscle and a kick of the foreleg. Similar myotatic reflexes can be elicited from muscle-tendon combinations anywhere in the body.

As Figure 4-1 shows diagrammatically, the knee jerk reflex is initiated by receptors in the muscle that are sensitive to stretch. These receptors generate impulses in sensory neurons which synapse on motor neurons in the cord that govern muscle contraction. The knee jerk reflex, it might be noted, is a segmental reflex. This means that the basic anatomical connections do not involve more than one segment of the spinal cord. In experimental animals, the neurophysiologist can

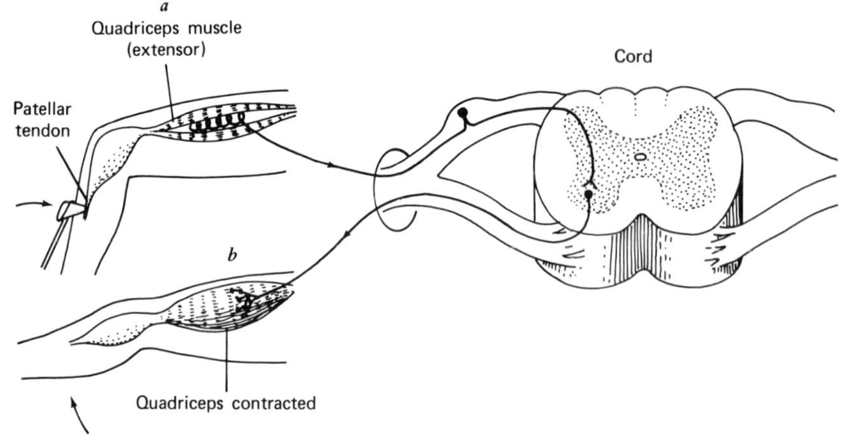

FIGURE 4-1. A diagrammatic representation of the knee jerk reflex.

surgically isolate a single segment of the cord in order to demonstrate that reflexes can take place on this level. Reflexes involving more than one segment of the cord, such as the automatic swinging of the arms while walking, are known as intersegmental reflexes and are mediated by the intersegmental pathways in the spinal cord.

SLOWER AND SMOOTHER: THE STRETCH REFLEX AND MUSCLE TONE

In principle, stretch reflexes involve the same elements as the simple knee jerk reflex. Both are myotatic reflexes; however, the stimuli that initiate them are different. The stretch reflex is a contraction of the extensor muscles in a limb in response to a gradual stretching caused either by the pull of gravity or by the contraction of antagonist muscles. A simple example of a stretch reflex occurs when we shift our weight from one leg to another. If the weight is transferred from the left leg to the right, the increased load imposed on the right leg causes the extensors of that leg to contract. If this failed to take place, the individual would collapse, the knee having buckled under. Because the neurons that fire in response to a slow stretch of the muscles do so irregularly and in gradually increasing numbers, the action is a smooth one in contrast to the rapid, total response typical of the knee jerk reflex.

Muscle tone describes the fact that some motor units in muscles are in a more or less constant state of slight contraction. Because of this, muscles do not start their contractions from zero, but instead have a head start. This lends smoothness to muscular reactions, particularly to those involving postural adjustments. Muscles lose their tonus if either the dorsal or ventral spinal roots are cut, demonstrating that muscle tone is mediated reflexly by sensorimotor arcs throughout the spinal cord.

SPINDLES, CHAINS, AND BAGS: STRUCTURES MEDIATING THE STRETCH REFLEX

Muscle contractions are initiated by the lower motor neurons, but as our discussion of reflex activity and muscle tonus indicates, there are complex mechanisms within the muscles and tendons themselves which respond to contraction or stretching that make such reflex activity possible. Returning to the knee jerk reflex, two types of receptors are present to signal the stretching of the muscle and its attached tendon. These are Golgi tendon organs (Fig. 4-2) and muscle spindles (Fig. 4-3). The Golgi tendon organs are embedded in the tendonous tissue that attaches muscles to bones. Because of this, Golgi tendon organs react whenever the tendon is stretched—whether by the contraction of the attached muscle or

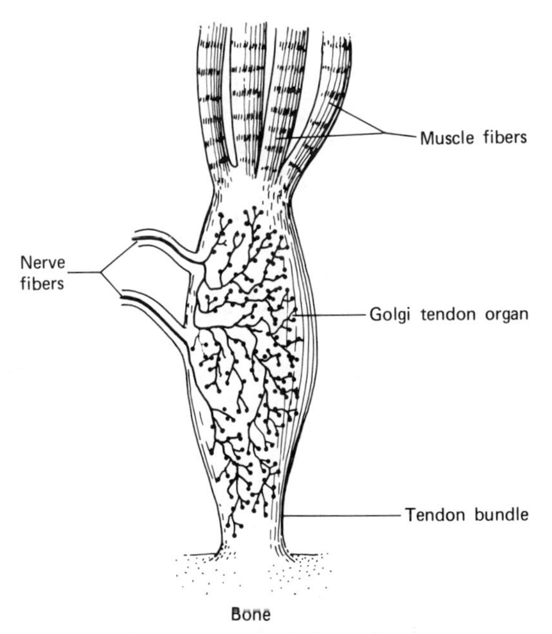

FIGURE 4-2. The Golgi tendon organ.

by the stretching of the muscle by the antagonist. Upon being activated, the tendon organs give a burst of impulses signalling the change. They then become quiescent even though the tendon continues to be stretched. These organs, therefore, provide information about changes in tension and not about the final position of the muscle-tendon complex. It is important to note that the sensitivity of the Golgi tendon organs is much less (perhaps 20 times less) than that of the muscle spindles that we will discuss next. Otherwise, the two would interfere with each other.

The second type of receptor is the muscle spindle shown diagrammatically in Figure 4-3. Muscle spindles are composed of neurons and muscle fibers that are enclosed in fusiform lymph-filled sacs or capsules (fusiform means rounded and tapering toward the ends). Because of this arrangement, the fibers are referred to as intrafusal (*intra,* within) in contrast to the extrafusal fibers to which the muscle spindles are attached at both ends.

Figure 4-3 also shows that there are two types of intrafusal muscle fibers, one whose cell nuclei are clustered in the center of the spindle to form a pouchlike structure called the nuclear bag. The second type contains nuclei that are stretched out in a chainlike fashion and are therefore called nuclear chain spindles. Both the nuclear bag and the nuclear chain spindles are surrounded by a high velocity type of sensory neuron called the Group I A or primary afferent. The nuclear chain fibers are also supplied by Group II or secondary afferents, which transmit lower velocity impulses. The Group I A or primary afferents inform the cord of both the velocity and extent of change in muscle length while the muscle is contracting. When muscle length has stopped changing, the response, which is called a dynamic response of the primary endings, sharply decreases. The secondary or Group II fiber also provides information about the length of the muscle, but the response here is called a static response, since it occurs when muscle length changes but continues its response as long as the muscle remains contracted.

The Group I A fibers are directly connected to extrafusal muscle fibers through synapses with alpha motor neurons that are the largest of the anterior horn cells supplying the muscles. Consequently, any increase in Group I A activity brings about increased contraction in the muscles. This relationship is an example of a positive feedback system. However, if there were no control over a positive feedback system, it would destroy itself. An analogy can be drawn here between an uncontrolled positive feedback system and a defective home furnace thermostat that calls for more and more heat the warmer the house becomes. Such a system would soon require the attention of the local fire depart-

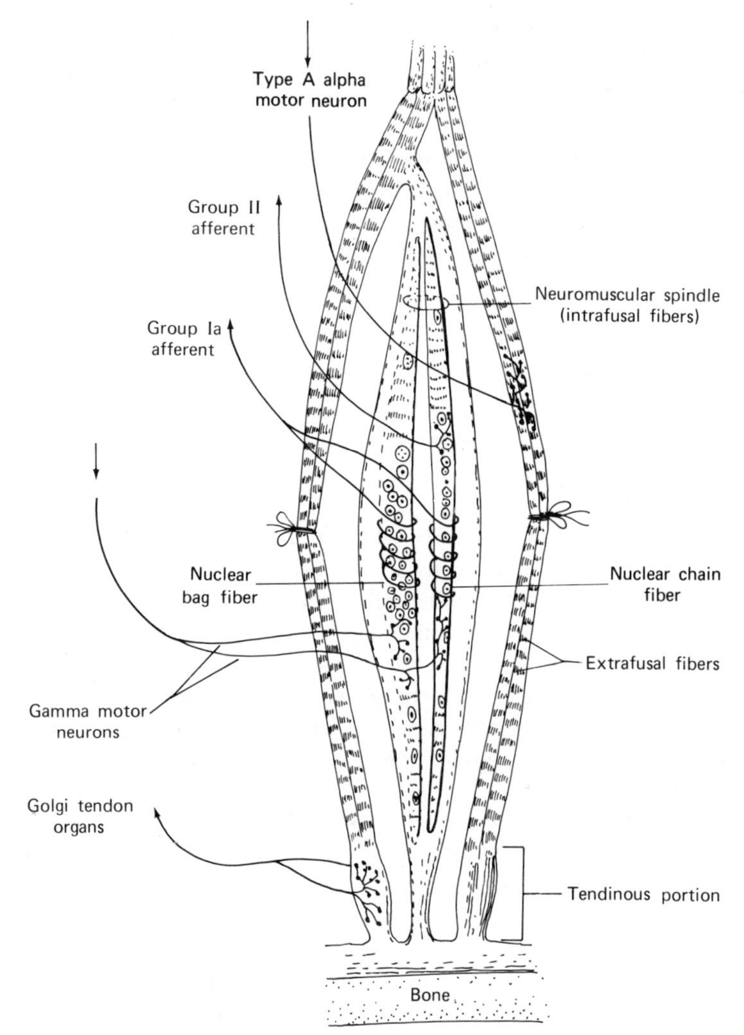

FIGURE 4-3. The muscle spindle.

ment. In actuality, of course, home thermostats are negative feedback systems, and the warmer the house becomes, the less heat is called for.

The muscle spindle system also has negative feedback components. Firing of the lower motor neurons activates Renshaw cells that synapse with collateral branches of the motor neurons. The Renshaw cells are, in turn, connected to adjacent motor neurons that they are capable of inhibiting (Fig. 4-4). The Group I A and II afferents also affect the balance of positive and negative feedback in complex ways depending on the circumstances of stimulation load imposed on the muscle, and on the degree of contraction in order to smooth out muscle response and prevent jerky contraction. We need not enter into the details of this process here.

Voluntary movement is superimposed on muscle tone and the stretch reflex mechanisms by large type A alpha motor neurons that are activated, in turn, by upper motor neurons of the corticospinal tracts. The gamma motor neurons—smaller neurons that are not involved in

EXPERIMENT IV

SHERRINGTON INVESTIGATES
RECIPROCAL INHIBITION

Sir Charles Sherrington (1857–1952), the distinguished British physiologist, contributed to almost every phase of neurophysiology by investigating decerebrate rigidity, all types of reflexes, the functions of the cerebral cortex, and pathological conditions. The bibliography of his publications numbers over 300 articles and books. A large proportion of these deals with the mechanisms of reciprocal inhibition or the principle that the spinal cord is capable of inhibiting an ongoing reflex in favor of a second, dominant, reflex without the necessity of cortical intervention.

The illustration shows a graphic recording of a scratch reflex being interrupted by a flexion reflex, the latter being prepotent over the former. The animal, a dog in this case, is prepared by sectioning the spinal cord to prevent the effect of cortical influences on the reflex under study. The dog must be suspended in a harness and recording equipment connected to the leg muscles. When the animal's skin is stimulated by a weak electric current in the region of the shoulder, the hind limb begins a rhythmic scratching movement with the paw brought forward to the site of the stimulation. If a strong electric current is then applied to the paw of the limb engaged in scratching, the scratch reflex will be inhibited in favor of a flexor reflex—the flexor muscles contracting strongly in an attempt to remove the paw from the source of irritation.

As the illustration shows, a central mechanism in the spinal cord inhibits the scratch reflex when the flexor stimulus is applied and maintains that inhibition until the stimulus ceases. Sherrington did not discover the precise mechanism of reciprocal inhibition, but he did

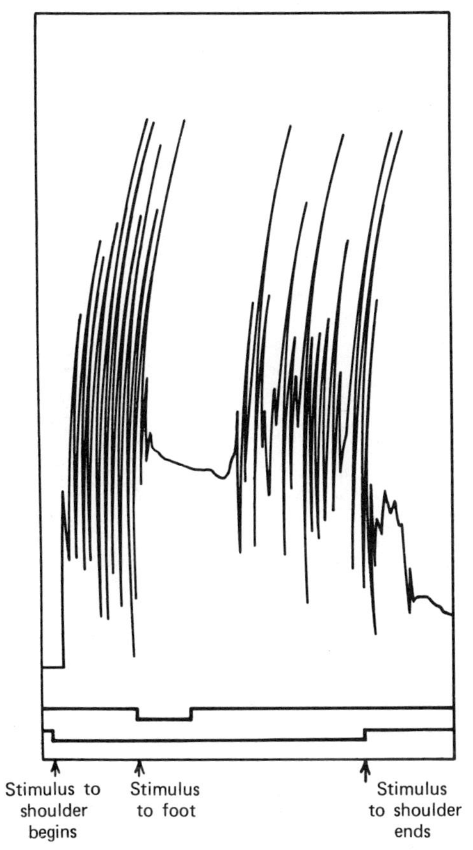

From Charles Sherrington, *Integrative Action of the Nervous System,* New Haven, Conn.: Yale University Press, 1948, Fig. 53, p. 192. Used by permission.

recognize it as a central process mediated at synapses in the cord. He called it the CIS or central inhibitory state. He also suggested that the process was probably chemical in nature.

It has since been discovered that there are two types of synapses, excitatory and inhibitory. Release of transmitter substances at the excitatory synapse produces a change in the cell membrane of the receiving neuron that is excitatory. Release at an inhibitory synapse produces a potential change in a receiving neuron that is opposite in polarity to the action potential and thus blocks transmission. It is to Sherrington's credit that this famous pioneer recognized the essential nature of the process over one-half century ago.

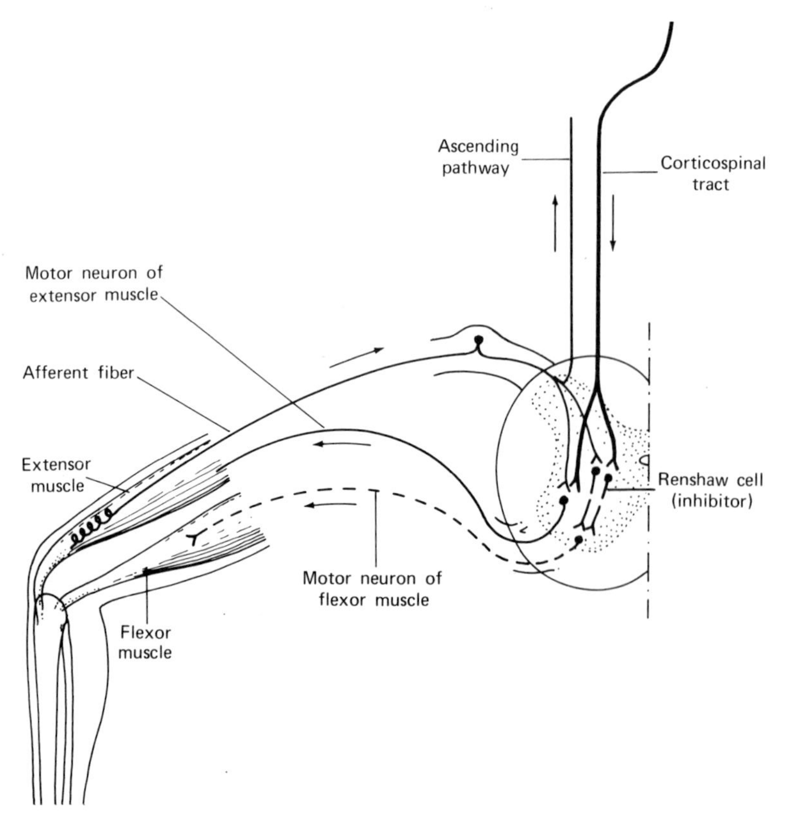

FIGURE 4-4. The inhibitory influence of the Renshaw cell. The stimulus from stretch to the extensor muscle is conveyed to the motor neuron for maintenance of tone of the extensor and to the Renshaw cell by way of a collateral synapse to inhibit the flexor muscle. These pathways are for control of tone. The descending pathway from the cerebral cortex is for control of voluntary movement of the muscle.

the feedback system—are also activated at the same time (see Fig. 4-3). These cause contraction of the intrafusal muscle fibers and temporarily increase activity from the muscle spindle I A fibers. When the muscle is fully stretched, the I A receptor activity decreases. To understand how this occurs, the sensory fibers wrapped around the muscle spindles may be thought of as analogous to a stretched spring when they are active and under tension. When the muscle is fully contracted, the tension on the springlike spindles is released, and firing of the I A sensory neurons decreases. In this way voluntary movement inhibits the stretch reflex, taking command over the muscles. Figure 4-4 summarizes the different feedback systems in reflex and voluntary control of muscular activity.

PROGRAMMING THE HOME RUN

We are now in a position to summarize our knowledge of the integrative action of the spinal cord as it is related to voluntary behavior. As a convenient example, we may use the baseball player at bat. He stands poised at the plate, his antigravity muscles supporting his body, his arms cocked at ready, holding the bat. His arms, like his legs, are kept up by antigravity muscles, the flexors in this case. Because human beings stand in an upright position, the antigravity muscles of the legs are extensors and of the arm flexors. In animals the antigravity muscles are all extensors.

All of our ballplayer's muscles are kept under tonus by continuous bursts of impulses transmitted by Group I A sensory fibers from muscle spindles that are responding to the stretching of the muscles because of the pull of gravity. To keep them from contracting too strongly, the Golgi tendon organs exert a constant negative feedback and balancing control over the motor neurons of the alpha system. At the same time, the player is conscious of the position of his joints and the pressures on them through the continuous firing of the Pacinian corpuscles, which are embedded in the tissues around the joints and which are highly sensitive to deep pressure.

When the ball is delivered and the player makes his split-second decision to strike at it, powerful contractions of the extensors in his arms are initiated by his corticospinal system. Because the extrafusal muscle fibers contract in his flexors, the muscle spindles shorten and their effects are inhibited. At the same time that these events are taking place, contributions from the cerebellum and vestibular centers are impinging on the final common path of the lower motor neurons.

When the batter strikes, the series of events that we have been describing is programmed to run off. He cannot make corrections in midswing once the bat has been swung even though his eyes tell him that he is going to miss before the umpire announces that disappointing fact. This suggests that the higher centers in the cerebral cortex initiate the program, but once underway the lower centers are responsible for its execution.

If our player hears the crack of the bat on the ball and feels the sharp change in the velocity of his arms, he knows that he has a hit. Immediately a complex set of postural and locomotor readjustments take place sending him on his way to first base.

Our seemingly detailed description is only a bare summary of what actually happens. So complex is the complete chain of neuromuscular events involved in the apparently simple act of hitting a baseball and

CASE IV

TABES DORSALIS

Mrs. T.P., a fifty-one-year-old widow, was first observed at the medical center hospital upon her arrival in the emergency room for treatment of an injured toe that had been fractured against a door jamb when she attempted to kick a dog and missed. A second significant incident occurred one year later when she slipped while walking, fracturing her ankle. In neither case was the patient observed extensively, having been treated by the application of casts in the emergency room.

Her current admission resulted from a history of further falls and mishaps. Her brother, who accompanied her, reported that "She falls backwards and forwards—never to any particular side." He also observed that his sister had shown changes in personality during the preceding two years with a tendency toward periods of depression, crying, and loss of interest in ordinary activities.

Mrs. T.P. was found to be well-oriented as to person and place, but deficient in memory for recent events. Language dysfunctions included inability to comprehend written material, disturbances in writing, and a tendency to employ neologisms with some aphasia, or difficulty in speaking.

Neurological tests of her limbs revealed hemiparesis or paralysis on the right side. Sensation was normal, with bilateral increase in muscle tonicity. Reflexes were hypertonic. The Babinski sign was present on the right side. The patient, upon questioning, admitted that she had experienced a loss in manual dexterity. In walking her gait was that of a tabetic. That is, she suffered from ataxia (*ataktos,* disordered) or incoordinated movements, slapping her feet down as if she were aiming for a stairstep that had suddenly been removed. This disturbance in locomotor behavior suggested that she was suffering from disease of the central nervous sytem, specifically the motor and kinesthetic pathways in the spinal cord with additional involvement of the reflex system. Further diagnostic procedures indicated the presence of an infectious disease, and a test for the presence of treponemal antigens revealed that she was suffering from tertiary syphilis.

The patient's symptoms persisted following antibiotic and physical therapy, because of irreversible neurological damage, but further deterioration was prevented, and the condition was described as stabilized.

making a run, they virtually defy description. A home run is worth cheering about, neurologically speaking.

SIGNS OF DISASTER: NOTES ON OTHER REFLEXES

Before leaving the subject of reflexes we would like to describe briefly several other patterns of reflex behavior that illustrate the integrative action of the spinal cord and are also useful in the clinic as diagnostic signs.

Perhaps the most famous of the diagnostic reflexes is that discovered by a Polish-born neurologist, Joseph Babinski (1857–1932), who practiced in France. Named after him, the Babinski reflex is elicited by stroking the sole of the foot. Normally the response to such stimulation is to flex all of the toes or to curl them downward. The Babinski sign, as it is called, is a dorsoflexion of the big toe (pulling it backward) and fanning out of the other toes. It is a sign of lesions of the upper motor neurons. Interestingly it also occurs in young infants for a few weeks after birth indicating that their cerebral cortices are not yet functional.

The so-called deep reflexes include the knee jerk and other myotatic reflexes associated with muscle-tendon groups. Neurologists routinely test for these reflexes to assist them in diagnosing possible organic damage to peripheral or central nervous pathways (see Case IV). Similarly, the pupillary reflex—the constriction of the pupil of the eye to light—is an important diagnostic sign. The absence of the deep tendon reflexes and of the pupillary response is typical of patients who are comatose because of neurological damage. The pupillary response, it should be noted, is mediated by motor nerves that arise directly from the brainstem rather than the spinal cord.

We shall return to the conductive and integrative functions of the spinal cord in Chapter 6 when we consider disorders of the cord. Meanwhile, in the chapter to follow we will describe a complex neural system that is closely associated with the spinal cord, the autonomic nervous system.

5
THE AUTONOMIC NERVOUS SYSTEM

GUT REACTIONS

In our descriptions of neurons, nerves, and spinal tracts we concerned ourselves primarily with the neural basis of somatic activity. In general, the somatic nervous system mediates voluntary activity—activity that is under the individual's conscious direction and is carried out by the skeletal muscles. But we also know there is another side to the machinery of the body, the visceral processes that are largely automatic and are not under our direct conscious control. The nervous system, which controls these functions, is appropriately named the autonomic nervous sytem, reflecting the fact that it has primary responsibility for regulating such visceral processes as blood pressure, secretions and motility of the gastrointestinal tract, respiration, output of the kidneys and sweat glands, and body temperature. While some of these processes may be consciously influenced by the individual in everyday living, they are basically automatic in nature, silently and efficiently conducting the business of the visceral organ system.

However, besides regulating the ordinary visceral functions of daily life, the autonomic nervous system is also intimately associated with emotional responses—the kinds of responses that we make in emergency situations or that develop during states of anger, joy, sorrow, or sexual excitement. The strong feelings we experience while undergoing these emotional reactions arise, in significant measure, from our awareness of the profound visceral responses that accompany emotion-provoking situations. In a sense, then, the autonomic nervous sytem mirrors the dualistic nature of living—the vegetative, concerned with the maintenance of normal bodily functions, and the emergency, concerned with crises or emotion-provoking situations. As we turn to its anatomy, we shall see that the divisions of the autonomic nervous sytem reflect this dualistic function.

ANATOMICAL OVERVIEW: A STRING OF BEADS

Anatomically the autonomic nervous sytem is a division of the peripheral nervous system, but unlike the fibers of the peripheral system that mediate both sensation from the receptors and control the voluntary muscles, autonomic fibers are entirely motor in function. Because the autonomic has no sensory or afferent fibers of its own, it utilizes those already available in the spinal nerves supplying the somatic and visceral receptors.

The autonomic nervous sytem has two divisions: the sympathetic, also known as the thoracolumbar; and the parasympathetic, also known as the craniosacral. The terms sympathetic and parasympathetic are more indicative of the functioning of these divisions than their anatomy. Sympathetic comes from the Greek *syn,* together, and *paschein,* to suffer or feel. The prefix *para* means beside or alongside of, expressing the idea that the parasympathetic division acts in a coordinate manner with the sympathetic division. The terms thoracolumbar and craniosacral are descriptive of the anatomical origin of the outflow of fibers of the subdivisions.

Figure 5-1 shows how the two systems originate anatomically. The sympathetic outflow arises from the spinal cord and a chain of ganglia that lie along either side of the spinal column (for simplicity only one chain is shown). The parasympathetic fibers arise from centers in the medulla and pons and from the sacral portions of the spinal cord. Each of the visceral organs receives fibers from the sympathetic ganglia, either directly or by way of a collateral ganglion. The same organs are also supplied with fibers from the parasympathetic division through long nerves that originate in the cranial region in the case of the thoracic and abdominal organs, or in the sacral centers of the spinal cord in the case of the pelvic organs. It should be noted that one organ, the adrenal gland, is innervated only by the sympathetic system. It has no parasym-

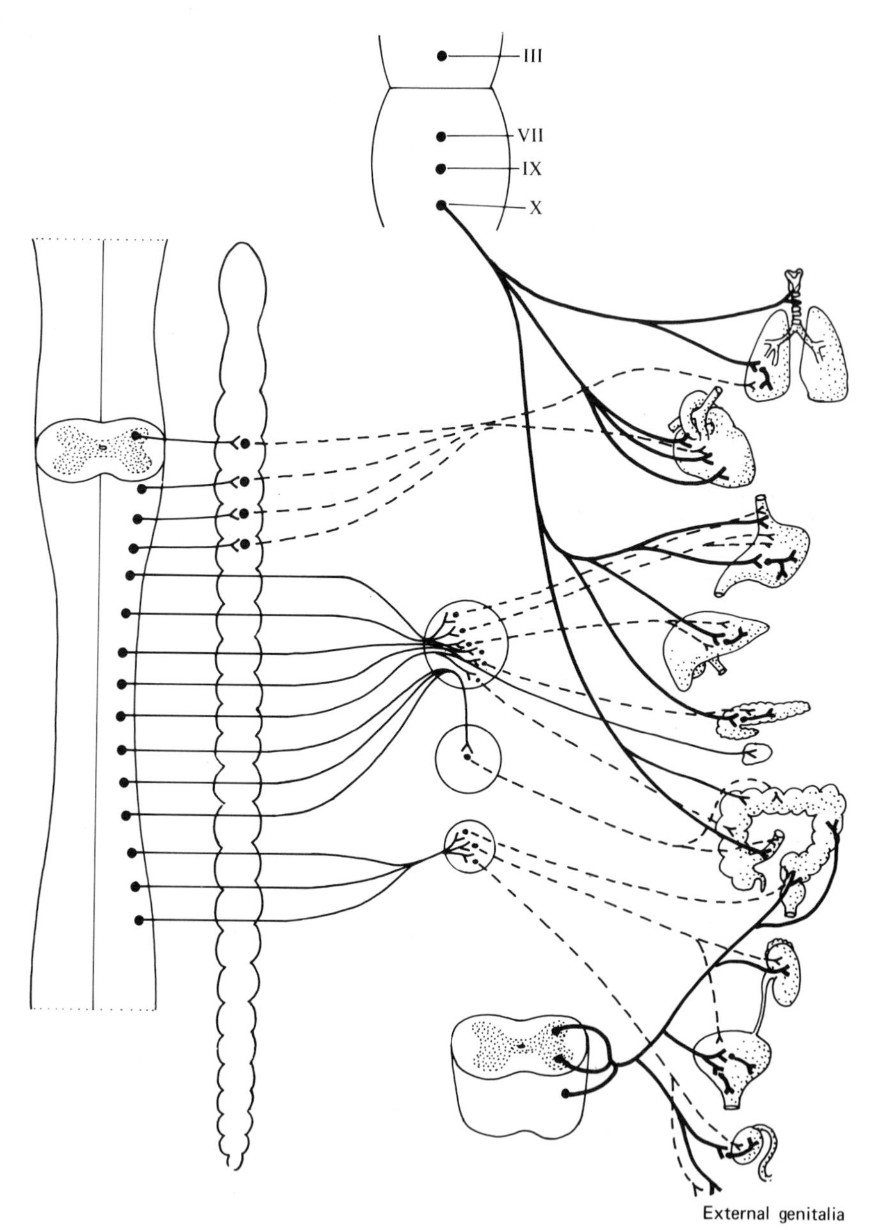

FIGURE 5-1. The origin and outflow of the sympathetic and parasympathetic division of the autonomic nervous system. Solid lines represent preganglionic fibers and broken lines, postganglionic fibers. Note that postganglionic parasympathetic fibers are close to or inside the organs innervated.

pathetic connections. The reasons for this arrangement will be made clear in a subsequent section.

In Figure 5-2 the close association of the sympathetic division with the spinal cord is shown in diagrammatic form. Functionally the relationship may be described as follows. Sensory neurons coming in over the dorsal roots of the spinal cord synapse with motor neurons of *both* the peripheral fibers that supply the voluntary muscles and the sympathetic fibers that innervate the visceral organs. On the left side of the

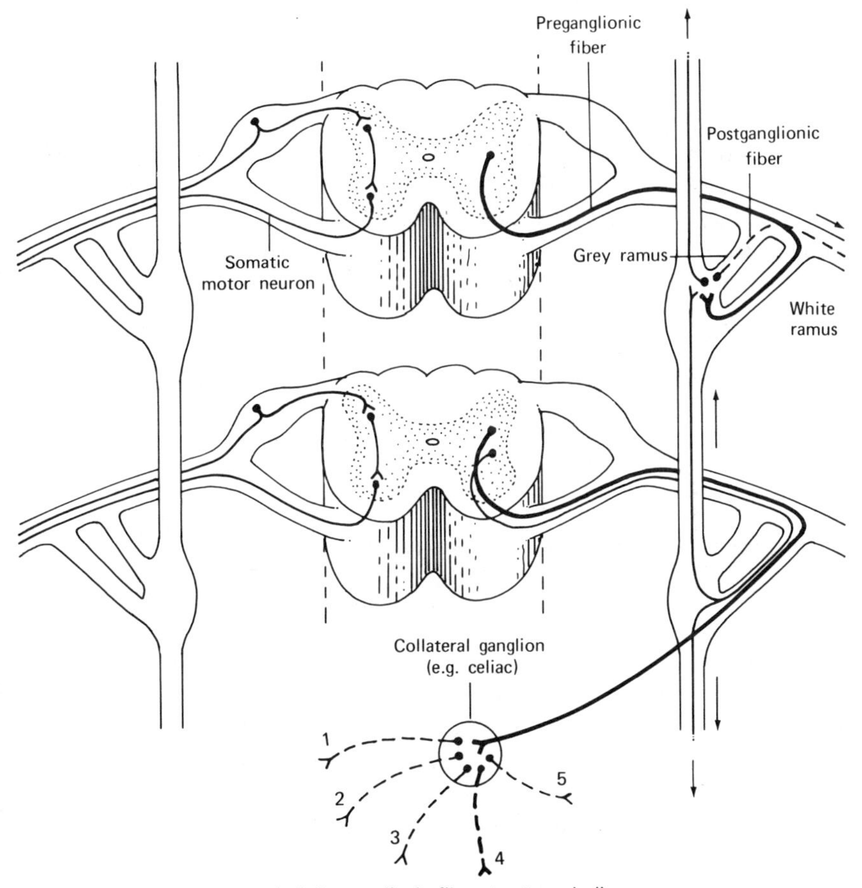

FIGURE 5-2. A schematic view of the connections of the sympathetic division of the autonomic nervous system with the spinal cord. Note that solid lines represent preganglionic fibers and broken lines represent postganglionic fibers. For explanation see text.

figure, a typical sensorimotor arc of the voluntary or somatic system is shown and on the right a reflex arc for the sympathetic system.

The sympathetic connections to the spinal cord are made through special trunks that attach the spinal nerves to the sympathetic ganglia. These trunks are called the gray and white rami (*ramus,* branch). The preganglionic neurons that originate in the spinal cord enter the ganglia of the sympathetic chain by way of the white rami. Here the preganglionic fibers either synapse immediately with postganglionic fibers or pass on through the chain of ganglia to a collateral or outlying ganglion, such as the celiac (Fig. 5-2). There they synapse with postganglionic fibers that supply visceral organs. The postganglionic neurons that originate in the autonomic ganglia and travel back out over the spinal nerves to the visceral organs are routed over the gray rami. The gray rami are bundles of unmyelinated neurons; the white rami are made up of myelinated fibers. The unmyelinated fibers of the autonomic are relatively small and slower acting as compared with those of the voluntary peripheral system. It might also be noted that the ratio of postganglionic neurons to preganglionic is high, making for a more diffuse response.

Some of the postganglionic fibers originating in the spinal ganglia pass into spinal nerves by way of the gray rami and travel out to the periphery of the body over the peripheral nerves of which they make up a significant portion. These fibers control the blood vessels, sweat glands, and piloerector muscles that make the hair "stand on end" or give the skin "goose bumps."

OF DRUGS: NATURE'S OWN AND MIMICS

The fibers of the two divisions of the autonomic nervous system are often spoken of as cholinergic and adrenergic, reflecting the fact that the terminals of the parasympathetic fibers liberate acetylcholine as their transmitter at the effector organs, while, with some exceptions, the sympathetic terminals release adrenalinlike substances (epinephrine and norepinephrine) and are therefore classified as adrenergic.

Certain drugs, because they cause the release of epinephrine and norepinephrine, are known as sympathomimetic. These include ephedrine (whose clinical usage is similar to epinephrine) and the amphetamines. Since these drugs bring about the release of stored sympathetic transmitters in synaptic endings, they mimic the action of general sympathetic discharge. Because they also affect the central nervous system, they may cause changes in behavior and consciousness. The effect of amphetamines (the "speed high") is a well-known example of such alterations.

Other drugs block adrenergic activity either by interfering with the synthesis and storage of norepinephrine in the sympathetic endings or by blocking its release. Reserpine, also employed as a tranquilizer, interferes with the synthesis of norepinephrine. Guanethidine, an important anti-high blood pressure agent, blocks the release of norepinephrine.

There are also drugs that act on the parasympathetic nervous system, and these are called parasympathomimetic (vagomimetic). Acetylcholine is the normal transmitter at postganglionic sites in the parasympathetic system. Pilocarpine and physostigmine, both parasympathomimetics, are used in the treatment of glaucoma because they constrict the pupil and widen the canal of Schlemm (see Case VI). Muscarine, a psychedelic drug, acts in ways similar to acetylcholine at postganglionic sites.

Two important agents, atropine and scopolamine, block the action of acetylcholine on effector organs. A common clinical use of atropine is the dilation of the pupil of the eye during ophthalmological examinations by blocking the action of acetylcholine on the sphincter muscle. Scopolamine depresses the reticular formation (see Chap. 11) causing a sleeplike state with amnesia—"twilight sleep" and for this reason is sometimes used in childbirth.

Finally, nicotine is a drug that can stimulate postganglionic neurons of *either* the sympathetic or parasympathetic. The result is a general discharge of both systems resulting in vasoconstriction in the viscera (sympathetic action) and increased gastrointestinal motility and slowing of the heart (parasympathetic effects). The sympathomimetic and parasympathomimetic drugs have found wide clinical usage in controlling a number of disorders whose course is influenced by autonomic effects.

RECPTORS: SECURITY MEASURES, LOCKS AND KEYS

Thus far we have identified the neurotransmitters of the autonomic nervous system, pointing out how drugs can either imitate or negate their effects. The scenario is incomplete, however, unless we consider the effector organs which, in a manner of speaking, are the receptors that interact with neurotransmitters. Table 5-1 reveals that blood vessels respond by either constriction or dilation "depending on receptors in

TABLE 5-1
Effects of Autonomic Stimulation on Selected Bodily Organs

Organ	Sympathetic Effects	Parasympathetic Effects
Eye		
Pupil	Dilation	Contraction
Ciliary process	None	Excitation
Gastrointestinal glands	Inhibition or no effect	Copious serous or watery secretion and enzymes
Salivary gland	Thick, viscous secretion	Serous or watery secretion
Sweat glands	Copious secretion	None
Heart	Increase in rate and force of contraction	Decrease in rate and force of contraction
Lungs	Constricts blood vessels, dilates bronchi	Constricts bronchi
Gastrointestinal muscle	Inhibits peristalsis, stimulates sphincters	Stimulates peristalsis, inhibits sphincters
Liver	Release of glucose	None
Penis	Ejaculation	Erection
Blood vessels	Constricts abdominal muscles, constricts or dilates other smooth muscles depending on receptors in tissue	None
Bladder	Uncertain	Stimulates smooth muscle for emptying, contracts detrusor, relaxes internal sphincter

tissues." In order to explain this and other contradictory physiological reactions, pharmacologists have developed a receptor theory that is akin to the complementary functions of locks and keys so familiar in everyday life. In essence, the theory suggests that the effector membrane has at its surface or contained within it a specific type of receptor or lock into which a neurotransmitter molecule or key fits. Basically, we are concerned here with the two types of adrenergic receptors: alpha adrenergic, which are excitatory (except in the gastrointestinal tract), and beta adrenergic, which are inhibitory (except in the heart). It has been postulated that noradrenalin stimulates ("unlocks") the alpha receptors and that adrenalin can cause a reaction *both* in alpha and in beta receptors when both are present within a given receptor. Therefore, adrenalin appears to be a kind of master key when both alpha and beta receptors are present in effector organs. The final result is the net difference between the two reactions or the sum of the two reactions. That is, an initial vasodilation (beta) of a vessel followed by vasoconstriction (alpha). The initial dilation is due to beta effectors' greater sensitivity to the transmitter; the second response is due to greater number of alpha receptors present in tissues.

BLUSHING OR BLANCHING: AN AUTONOMIC REFLEX

The simplest level of functioning of the autonomic nervous system can be exemplified by a reflex involving constriction or dilation of the blood vessels. The walls of blood vessels are made up of smooth muscle, and their reactions of constriction and dilatation are reflexively controlled. Let us assume that someone thrusts his hand into a basin of warm water. Sensory neurons immediately begin transmitting impulses to the spinal cord from thermal receptors in the tissues of the hand. In the cord these sensory neurons synapse with preganglionic sympathetic neurons that convey, in turn, impulses to the ganglia and by way of the gray ramus to the blood vessels. There the impulses cause dilation of the vessels in response to heat in an attempt to preserve thermal constancy in the limb. If an extremity becomes cold, the opposite reaction of vasoconstriction occurs in order to prevent excessive chilling of the blood.

Parasympathetic reflex control of specific organs is mediated by similar arcs involving sensory neurons from the peripheral system and components that arise from centers in the medulla-pons region or the sacral portion of the spinal cord (Fig. 5-3). Organs in the upper part of the body, such as the heart, stomach, and lungs arc served by motor fibers of

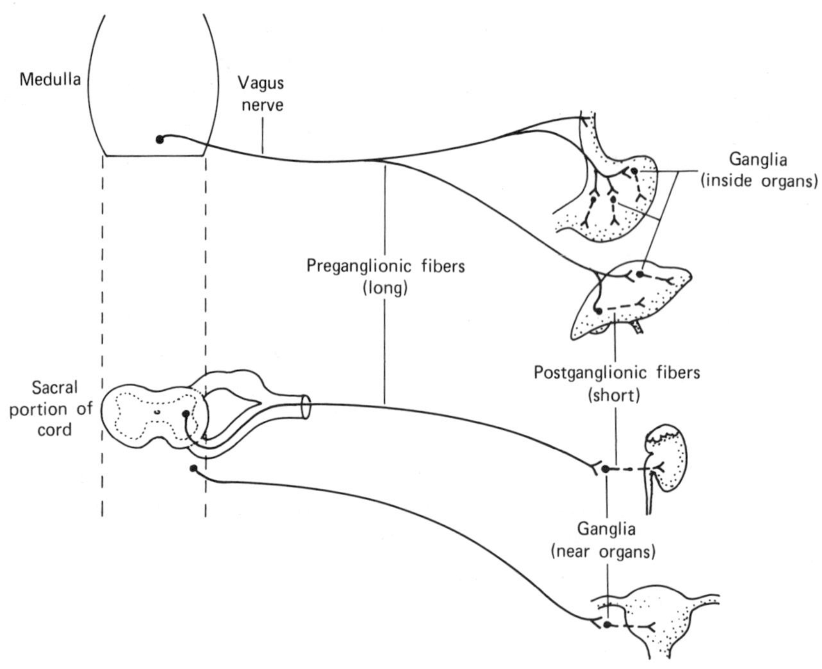

FIGURE 5-3. A schematic view of the connections of the parasympathetic division of the autonomic nervous system with the sacral portion of the spinal cord and the medulla. Note that solid lines represent preganglionic fibers and broken lines represent postganglionic fibers. For explanation see text.

the parasympathetic system which originate in nuclei in the brainstem and join one of the cranial nerves called the vagus (*vagus,* wandering, straying), appropriately named, since it wanders down through the thorax and upper abdomen giving off branches to the visceral organs. The parasympathetic neurons in the vagus nerve are preganglionic fibers. They synapse with short postganglionic neurons that are located near to or within the tissues of the organs themselves.

For the control of pelvic organs, the preganglionic parasympathetic fibers originate in the sacral portion of the spinal cord and form the pelvic nerve. This nerve supplies the colon, rectum, and sexual organs. To illustrate parasympathetic reflex control over the visceral organs and its close coordination with the activities of the sympathetic division, we will describe control of blood pressure, respiration, the bladder, and stomach since the functions of these organs often come to the attention of the clinician.

The control of blood pressure is accomplished through the coordi-

CASE V

SYMPATHECTOMY

At the time of his admission to the medical center hospital, H.T. was sixty years of age. He was employed as a receiving clerk in a warehouse—a job requiring him to stand for most of the day. Increasing episodes of pain in both of his lower limbs were experienced whenever he attempted to engage in extended periods of activity. Relief was experienced upon cessation of activity. A prior hospitalization three years earlier had been necessary for the treatment of high blood pressure and the onset of diabetes mellitus. At that time it was determined that he had a family history of a cardiovascular disease. In an attempt to correct his high blood pressure the patient was placed on vasodilating drug therapy.

Because the previously ordered treatment had failed to correct the condition, arteriography, or the examination of the blood vessels in the patient's legs by means of radiopaque dye, was undertaken in order to decide upon a course of further treatment. The results revealed narrowing along the midportion of the femoral artery with poor blood flow below the popliteal artery in the left leg. The tibioperoneal arterial supply to the lower portion of the leg was completely occluded. The patient's condition was considered severe enough to warrant surgical intervention in the form of a sympathectomy in order to delay or possibly avert the necessity of amputation of the limb.

A left lumbar sympathectomy was performed to dilatate the arteries and increase blood flow. The operative procedure was successful and the patient was able to return to his job. He was advised to stop smoking because of the vasoconstrictor effects of nicotine on the circulatory system and was also encouraged to develop collateral circulation in his legs by a program of special exercises.

nated action of several regulating mechanisms. First, on the periphery of the body, receptors in the skin and effector neurons supplying the blood vessels function through reflex action of the sympathetic to control blood flow as already described. However, it must be borne in mind that peripheral constriction will also increase vascular pressure, since the cardiovascular system is a closed one. Second, sets of receptors sensitive to changes in blood pressure are located in the aorta and in the common carotid arteries in the neck. Collectively, these are known as baroreceptors since they are pressure detectors. If blood pressure tends to fall to below average levels, these receptors send messages to the cardioaccelerator centers in the medulla, which in turn send impulses by way of the sympathetic ganglia to the heart causing it to beat faster. If pressure tends to rise significantly, this same reflex circuit inhibits the heart by way of the parasympathetic system. These mechanisms are shown diagrammatically in Figure 5-4. Average blood pressure is maintained by the delicate balance of antagonistic stimulation from sympathetic and parasympathetic centers.

Control of respiration is a joint function of somatic and visceral systems. The centers governing respiration receive impulses from higher centers making it possible for the individual to voluntarily modify

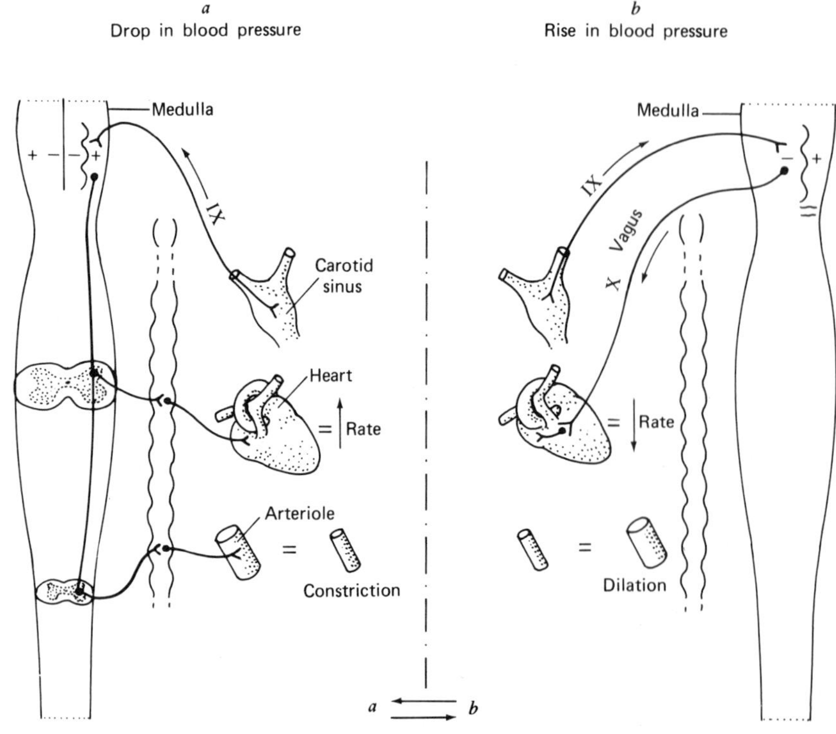

FIGURE 5-4. The autonomic control of blood pressure. Side *a* represents responses of baroreceptors to a drop in pressure. The vasomotor center in the medulla (+) triggers an increase in heart rate and vasoconstriction by way of the ninth cranial nerve. Side *b* represents responses of baroreceptors to an increase in pressure. The vasomotor center in the medulla (−) is inhibited with the result that the parasympathetic dominates by way of the vagus nerve to decrease heart rate and allow reflex dilatation.

his respiratory movements in such activities as speaking, singing, playing musical instruments, and the like. The muscles controlling these movements in the diaphragm are voluntary in type. The anatomy and physiology of these centers will be considered in more detail in Chapter 9.

However, basically respiration is controlled reflexively by visceral mechanisms. This is demonstrated by the fact that no one can asphixiate himself by voluntarily holding his breath. Figure 5-5 shows the centers in the pons and medulla, which regulate inspiration and expiration. Here we may note that the centers in the medulla controlling inspiration send

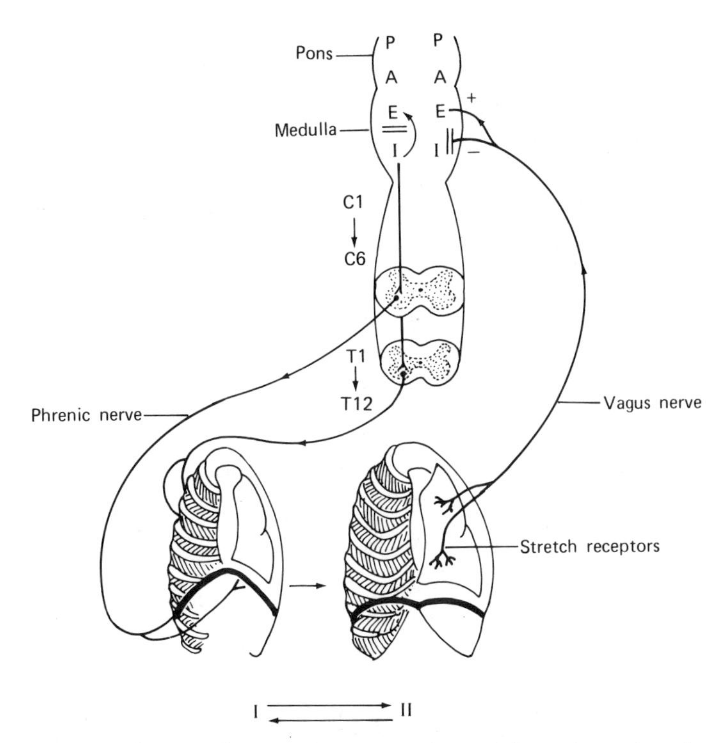

FIGURE 5-5. A schematic diagram of centers in the medulla and spinal cord regulating respiration (A = apneustic; P = pneumotaxic). Inspiratory cells in the medulla send impulses down the cord to the intercostal muscles via the intercostal nerves and to the diaphragm via the phrenic nerve (left side of diagram). Note that inhibitory impulses are simultaneously sent to the expiratory center (E) by these inspiratory cells. This leads to expansion of the thorax (right side of diagram). Progressive stretch of the receptors in the alveoli of the lungs activates the vagus nerve, which inhibits the inspiratory center (I−) and excites the expiratory center (E+). In summary, it should be noted that when one center in the medulla is active, the other is inactive. The influence of upper centers in the pons is considered in Chapter 9 (see Fig. 9-7).

impulses over the phrenic nerve, which cause the contraction of the muscles of the diaphragm. Other centers in the same general area are capable of inhibiting the inspiration centers and so allowing for expiration to occur. However, there are also receptors within the lungs themselves that are sensitive to the stretching caused by inspiration. These receptors are connected to the pons and medulla by fibers associated with the vagus nerve. When the frequency of discharge of the fibers becomes sufficiently high, the inspiratory centers are inhibited and expiration occurs. Lesions of the spinal cord above the fifth cervical segment interrupt these pathways causing rapid death due to failure of respiration. Finally, the gaseous condition of the blood also exerts control over the respiratory centers by way of special detectors or chemoreceptors in the carotid arteries and the aorta. Until the discovery of the Salk poliomyelitis vaccine, a number of individuals died each year if they were not artificially maintained in iron lungs (tank-type respirators) as a result of viral infections of the respiratory centers in the medulla.

The sympathetic neurons innervating the bladder originate in the lumbar region of the spinal cord, pass to a collateral ganglion, and synapse with a postganglionic neuron that terminates in the muscular walls of the bladder and the internal sphincter (Fig. 5-6). Surprisingly,

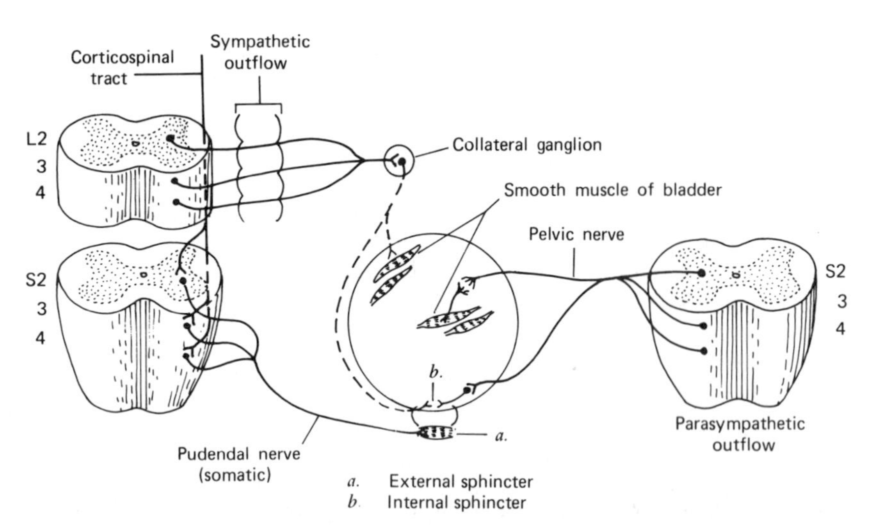

FIGURE 5-6. The innervation of the bladder showing the pudendal and pelvic nerve. The pudendal nerve, which is a part of the somatic system, controls the external sphincter (*a*) while the pelvic nerve is a part of the parasympathetic system controlling relaxation of the internal sphincter (*b*) and contraction of the bladder wall.

the functioning of the sympathetic division of the autonomic nervous system in bladder control is not yet well understood. From experimental studies of animals, we know that sympathetic stimulation may tend to relax the muscles in the bladder wall allowing for filling, but has little or no effect on emptying the bladder, as demonstrated by the absence of any significant change in the capacity to initiate urination after sympathetic fibers are cut.

Parasympathetic innervation of the bladder takes place through the pelvic and pudendal nerves. The preganglionic fibers for these nerves originate in the gray matter of the sacral region of the spinal cord. The pelvic nerve supplies the muscular walls of the bladder and the internal sphincter. The pudendal nerve, a spinal nerve under voluntary control, supplies the muscle of the external sphincter. Parasympathetic stimulation contracts the muscles in the wall of the bladder and relaxes the internal sphincter. The external sphincter is made of striated muscle that may be voluntarily controlled through the pudendal nerve. However, this sphincter relaxes reflexively when urine begins to flow through the internal sphincter.

Parasympathetic innervation of the stomach and intestines through the vagus nerve stimulates digestion by increasing peristalsis and the secretion of digestive juices (Fig. 5-7). Sympathetic innervation, mainly through the splanchnic nerve, tends to inhibit these functions. However, chronic stress or emotional disturbances may cause parasympathetic overcompensation with the result that peristalsis and the secretion of hydrochloric acid is overstimulated causing the mucosal lining of the stomach or intestine (usually the first few inches of the duodenum or portion of the small intestine leading off the stomach) to break down and an ulcer is produced. In order to control stress, a person may be given tranquilizing drugs, and to inhibit overactivity of the stomach and intestines, agents that block parasympathetic activity. Antacids may also be recommended to neutralize some of the excessive acidity of the stomach.

The examples that we have given involve stimulus-response reflex operations of the parasympathetic and sympathetic divisions of the autonomic nervous system. However, the reactions of the system are often far more widespread than our examples suggest. Table 5-1 summarizes autonomic effects on various bodily organs. In general, it may be noted that the two systems balance or complement each other. Stimulation by the sympathetic inhibits some organs and excites others. Conversely, stimulation by the parasympathetic inhibits or excites, depending on the organ involved. For the most part, an organ (except for organs such as the stomach in which the sphincter may be controlled by one division

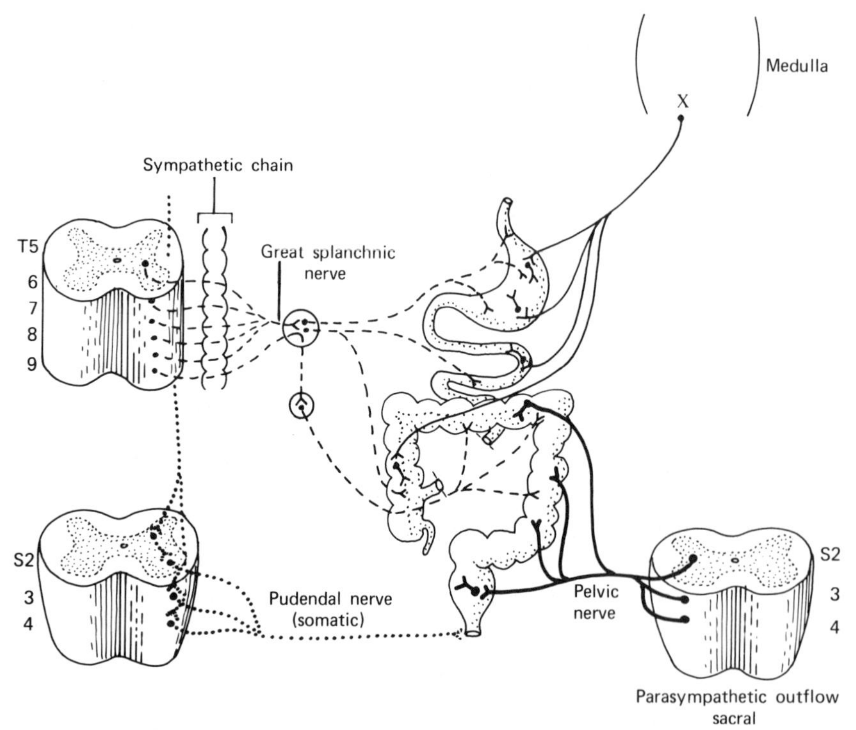

FIGURE 5-7. The innervation of the gastrointestinal tract. The parasympathetic system shown in solid lines (pelvic nerve) stimulates digestion. The sympathetic system shown in broken lines inhibits digestion. Control of the external sphincter is by way of the somatic system (pudendal nerve) shown in dotted lines.

while the organ itself is controlled by the other division) is either under the influence of one or the other of the systems at any given time, thus avoiding antagonistic responses. It is also important to note that the parasympathetic and sympathetic systems, like the skeletal system, maintain a certain degree of tone over organs. Thus, blood vessels and the smooth muscles of the gastrointestinal tract are in a state of partial constriction or contraction. This arrangement allows for more sensitive control over the visceral organs by the autonomic. Moreover, as everyday experiences involving fear, love, or anger show, emotional situations stimulating the higher centers of the central nervous system are capable of triggering visceral reactions under the control of the autonomic nervous system. To these more general functions of the autonomic we shall now turn.

FIGHT OR FLIGHT: GENERAL AUTONOMIC FUNCTIONS

Broadly speaking, the functions of the sympathetic division have been described by Walter B. Cannon (1871–1945), a famous Harvard physiologist, as emergency reactions for fight or flight (Experiment V). As we have seen, when dominant, the sympathetic system diverts blood from the digestive organs to the voluntary muscle system, thus inhibiting digestion. Salivation, another digestive function, is also inhibited. The cardiac and respiratory rates are increased in order to provide more oxygen for the muscles. Sweating is stimulated to keep body temperature from rising. These effects are brought about by the secretion of epinephrine or adrenalin at sympathetic endings in the organs and by direct discharge of epinephrine into the bloodstream by the adrenal medulla whose secretory cells are modified sympathetic tissue. It is interesting to note that the adrenal glands receive special preganglionic sympathetic fibers from the intermediolateral matter of the spinal cord. These pass without synapsing directly to the adrenal medulla where they synapse on the secretory cells that produce epinephrine. Because the pattern of postganglionic neurons of the sympathetic nervous system is widely dispersed and because the stimulation of the visceral organs is reinforced by the discharge of epinephrine, the emergency reaction is a diffuse one that is slow to disappear.

As predicted by the emergency theory, parasympathetic reactions would, in general, be the opposite of sympathetic reactions, and this is mainly what is found. The heart is slowed, breathing rate decreases, blood is diverted to the digestive functions and sweating is inhibited. However, parasympathetic reactions are more discrete or specific in their effects than sympathetic. As a review of Figure 5-1 will reveal, the postganglionic neurons of the parasympathetic nervous system are located close to or within the organs themselves in contrast to the postganglionic neurons of the sympathetic system, which arise from the central control centers near the spinal cord. Moreover, as we have noted, the chemical transmitter of the parasympathetic division is acetylcholine, which is secreted by the postganglionic synapses directly into the organ, a chemical transmitter unlike epinephrine, which is discharged not only at postganglionic synapses but also directly into the bloodstream.

Because emergencies are atypical situations, it is best to think of the two divisions of the autonomic nervous system as complementary rather than antagonistic in function. Some degree of diffuse sympathetic reactions, it is true, occurs in response to every stimulus; but by means of a

EXPERIMENT V

BARD INVESTIGATES THE NATURE OF THE RAGE RESPONSE

Walter B. Cannon and Philip Bard of Harvard University developed their famous emergency theory of the emotions on the basis of a long series of investigations on autonomic and central mechanisms that underlie the expression of the emotions in animals. One of the pivotal or key reports on the neural processes underlying emotional behavior was Bard's 1928 paper, "A diencephalic mechanism for the expression of rage with special reference to the sympathetic nervous system," in the *American Journal of Physiology*.

In his report Bard summarizes experiments on 46 cats, which involved the destruction of the cerebral cortices followed by removal of various portions of the remainder of the brain. The purpose of these experiments was to determine the precise centers for the control of the various phases of the rage reaction.

The results of these experiments confirmed Cannon's earlier findings that animals whose cortices have been destroyed show "sham rage," which develops either spontaneously or in response to the slightest provocation. Bard describes the response as follows:

> struggling, attended by movements of the head and arching of the trunk with thrusting and pulling of the limbs; clawing movements of the fore legs with protrusion of the claws; waving and lashing of the tail; a snarling expression; and very rapid panting with mouth open and movements of the tongue to and fro. In addition to these activities were signs denoting a vigorous sympathetic discharge; erection of the tail hairs; sweating from toe pads; retraction of the nictitating membranes; exopthalmos (separation of the lids); large increments in arterial pressure and heart rate. (From Bard, *American Journal of Physiology*, 1928, p. 494.)

Such reactions are called "sham rage" since they are not directed at any outside source of threat to the animal and, unlike normal rage responses, are likely to disappear rapidly. However, they indicate that centers below the level of the cerebral cortex do mediate the visceral changes in rage.

Sectioning of the brain of decorticated animals at various subcortical levels, both anterior and posterior to the region of the hypothalamus, demonstrated that the hypothalamic region was responsible for mediating the sympathetic activities involved in the rage response. Animals whose hypothalamus was sectioned leaving only the medulla and cord intact were incapable of sham rage. These and other observations by Bard, Cannon, and others revealed three general levels of neural functioning in emotional behavior: the cortical, subcortical, and autonomic. The cortex acts in the perception of emotion-provoking situations and directs and prolongs the animal's response. The hypothalamic and associated subcortical muclei in the brain act as integrating and control centers over the sympathetic nervous system. Finally, the sympathetic nervous system provides the motor outflow that generates what Cannon called the emergency reaction typical of rage or anger.

series of discrete or specific compensatory reactions, the parasympathetic maintains a balance. Cannon believed this process was part of homeostasis (*homo,* same; *stasis,* state), or the maintenance of a steady state or constant internal environment—a process that he believed was the function of bodily reactions in general but particularly of those involved in temperature regulation, maintenance of blood sugar level, and cellular metabolism.

FROM ON HIGH: A NOTE ON CEREBRAL INFLUENCES ON AUTONOMIC FUNCTIONS

While the autonomic nervous system is basically a peripheral system, it is not independent of control by higher centers in the brain. We shall be considering these influences in detail in Chapter 13, but we would like to point out here that the hypothalamus has been described as "the center of the emotions." Whether or not it deserves so exalted a title, there is no doubt that it is one important center of emotional integration and arousal. Its destruction in animals causes a loss of integrated emotional behavior. Its stimulation arouses fear, rage, appetite, or sexual behavior, depending on the site stimulated. As we know from everyday experience, these emotional states are correlated with profound and widespread changes in visceral reactions, thus demonstrating the close connection between the hypothalamus and the autonomic system.

The cerebral cortex also plays a significant part in integrating and sustaining emotional behavior. Animals whose cortices have been removed appear to be incapable of sustaining and directing emotional reactions. We know, too, that imagining or thinking about emotional situations such as love, anger, sex, or grief can arouse and sustain emotions. We shall have more to say about the cerebral control of feeling and emotion in Chapter 13.

6

SPINAL INJURIES AND DISEASES IN MAN

LEARNING FROM DISASTER

Diseases or injuries to the spinal cord can be catastrophic for the victim, leaving him without sensation and either totally or partially paralyzed below the level of the lesion. The inability of spinal neurons to regenerate following injury means that such injuries are irreversible. And yet, because of the architecture of the spinal cord with its alternate and crossed pathways, functions may not be lost completely in injury and disease except, of course, when the cord is completely transected. And even though clinicians cannot restore tissues that have been destroyed, they can assist the individual to cope with his disability by teaching him to make the best possible use of what function remains.

Clinicians are not the only specialists interested in diseases and injuries to the spinal cord. Anatomists and neurophysiologists have found these disorders to be their single most important source of knowledge about the conduction functions of the cord in man. In order to identify the pathways involved in cases of spinal injury or disease, behavioral disabilities are correlated with post-mortem pathological examination. This, in fact, is primarily how we know which pathways are responsible for the many sensory and motor functions of the spinal cord. We shall consider some of the more common injuries and diseases that man is heir to and that have proved a challenge to the clinician or have provided the anatomist and neurophysiologist with knowledge of the cord's structure and functions.

TOTAL DISASTER: TRANSECTION OF THE CORD

Complete transection of the spinal cord results in a clinical syndrome (*syn*, with, together; *dramein*, to run) known as spinal man. If the transection occurs above the fourth cervical vertebra, death follows immediately because of interruption of respiratory pathways, unless, of course, artificial respiration is maintained. Transection of the cord at lower levels causes complete paralysis and anesthesia of all body segments below the level of transection. Transection is also accompanied by the temporary suppression of all reflex activity. This condition is called spinal shock. Reflex activity gradually returns in one to six weeks following the injury, signalling the end of spinal shock.

The first sign of returning reflex function is the appearance of the flexion reflex in response to intense stimulation of the sole of the foot. This flexor activity gradually becomes more complex with withdrawal of the limb accompanied by flexion at the hip, knee, and ankle. The Babinski sign is present. It should be emphasized that the subject does not consciously experience the sensation of being stimulated, nor is he directly aware of muscular movements of the limb that is responding. If blindfolded, he would be unaware of both the stimulation and his response to it. Autonomic fibers are also involved in the flexor response as indicated by sweating and emptying of the bladder and rectum upon cutaneous stimulation of the feet, legs, or thighs.

Because of the widespread nature of these responses, the flexion reflex following spinal shock is known as the mass reflex, or mass flexion. The initial suppression of voluntary and autonomic reflexes is believed to be the result of the sudden withdrawal of control from the higher centers, since in experimental animals gradual destruction of the spinal cord does not leave the animal in a state of spinal shock. The same finding has been observed in human subjects whose lesions are the result of slowly developing tumors.

Voluntary control of the muscles is not reestablished following the disappearance of spinal shock. Because the lower motor neurons are not stimulated by higher centers, the muscles below the level of the lesion remain in a state of paralysis and eventually atrophy. Clinically such persons are known as paraplegics (*para*, beyond; *plessein*, to smite) if the lower limbs and body are involved and as quadraplegics if all four limbs are involved.

Following spinal transection, autonomic functions are at first completely suppressed. The bladder and rectum are paralyzed, and urination and defecation must be induced by catheterization and the use of enemas. As shock wears off, the bladder tends to become incontinent, with spontanious emptying of the filled bladder, a condition known as spinal bladder. Similarly, return of rectal function with involuntary defecation occurs three or four weeks after transection as shock subsides. When bladder and rectal functions have been restored, the mass reflex can be used to trigger evacuation of urine and feces. However, the usefulness of the technique depends on teaching the patient to follow a regimen, otherwise such reflexes may be initiated spontaneously at inappropriate times and places without specific stimulation.

Sexual functions are seriously disrupted, primarily because of loss of sensation from the genital organs. In spinal men, genital stimulation will produce erection after recovery from spinal shock. Ejaculation of semen may also be induced by manual or electrical stimulation. These patients, however, feel nothing, and if prevented from seeing what is taking place, are unaware of either erection or ejaculation. A small percentage of spinal men are capable of sexual intercourse and of impregnating women. In such cases, women must stimulate the men manually or orally and must, of course, take complete initiative in carrying out coital activities. Following spinal transection, few men have reported some residual sexual desire. However, in these individuals psychological factors are responsible rather than sensations arising from the genital organs.

A STAB IN THE BACK: THE BROWN-SÉQUARD SYNDROME

Occasionally the spinal cord is hemisected, or cut in half, by a stab wound or similar injury. The French-American neurologist, C.E. Brown-Séquard (1817–1894), for whom the syndrome is named, observed that in such cases spastic paralysis occurs on the same side of the body while sensation is interfered with on the side opposite the lesion.

A review of the conduction pathways will show why this is so (Fig. 6-1). The majority of fibers for the motor pathways cross in the medulla before descending. Consequently, interrupting these fibers affects the same side. Because the neurons for pain and temperature cross shortly after entering the cord, these modalities will be affected on the opposite side. Touch is only slightly impaired because of the alternate pathways

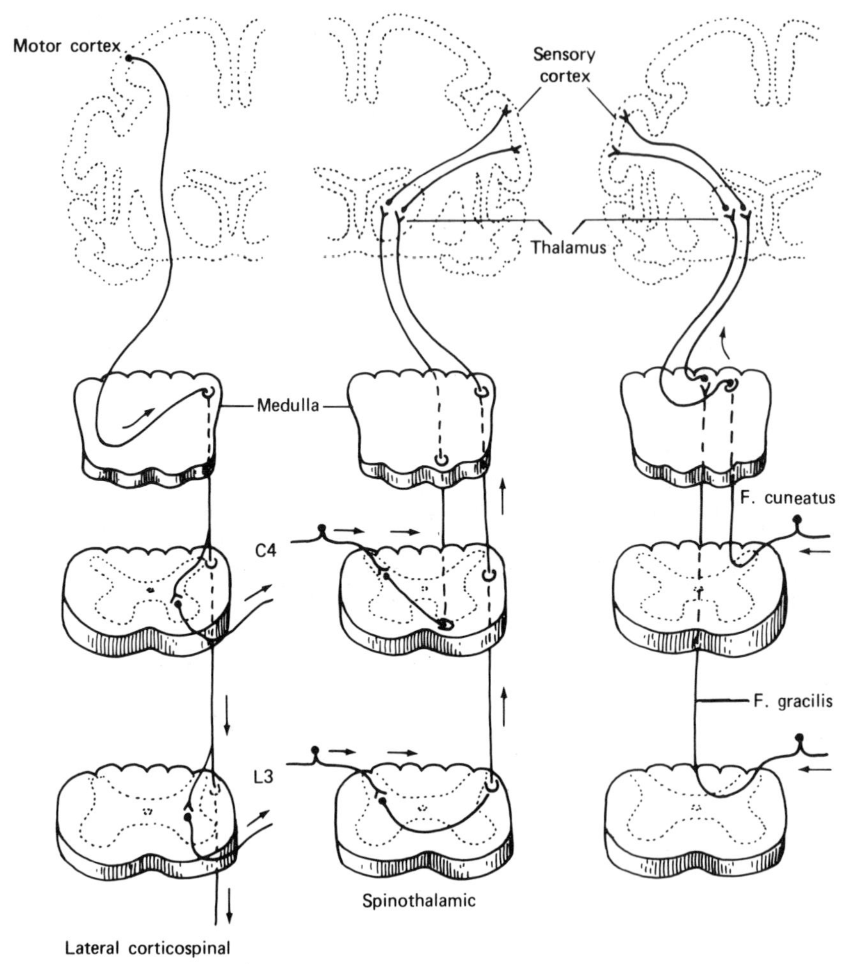

FIGURE 6-1. Principal tracts affected by hemisection of the spinal cord.

that are available. Some kinesthetic loss and impairment of pressure sensitivity may occur as a result of the interruption of fibers ascending on the same side by way of the fasciculus gracilis and fasciculus cuneatus, which do not cross until they reach the level of the medulla. However, there are alternate pathways for kinesthesis, and this sense modality is not seriously disrupted.

Because of the complexity of spinal pathways, some of which are still imcompletely understood, the symptoms of hemisection of the cord are variable from individual to individual. Moreover, it is unusual for any one sense to be completely eliminated on the side opposite the hemisection, apparently because of the availability of alternate pathways.

WITHOUT PAIN: NOT A BLESSING

Syringomyelia (*syrin,* tube; *myelos,* marrow) as its derivation from Greek implies, is a condition of the spinal cord in which the central canal enlarges into a hollow cavity with destruction of surrounding tissues (Fig. 6-2). The cavitation may be caused by one of several disorders, including congenital malformations or degenerative changes following hemorrhages or neoplasms. The cavitation may extend for six or more segments of the cord. Because it originates in the central canal and spreads outward, the disorder first involves immediately surrounding tissue. For this reason the initial symptoms are loss of pain and temperature because the fibers for these sense modalities cross the cord by way of the anterior commissure; because of this crossing, the opposite side of the body below the lesion is affected. However, syringomyelia is a slow but progressive disease that extends both outward into the surrounding tissues of the spinal cord as well as vertically up and down the cord. These progressive changes gradually involve other sensory and motor modalities with touch, kinesthesis, and voluntary and reflex control of the muscles, all being eventually impaired.

Since we are discussing pain, we would like to take the opportunity to point out that occasionally an individual who, because of a congenital deficiency in the central nervous system, is born without a sense of pain. Such a condition might, at first thought, appear to be advantageous. On the contrary, these individuals rarely enjoy a full life-span because of constant injuries of which they are unaware or because of disease symptoms, normally signalled by pain, which go unnoticed and prove fatal. These people must learn to use other cues to alert themselves to dangers in the environment and to the presence of disease, but even with careful observation, they are likely to overlook serious burns, bruises, fractures, and disease processes. A life without pain may be comfortable but short.

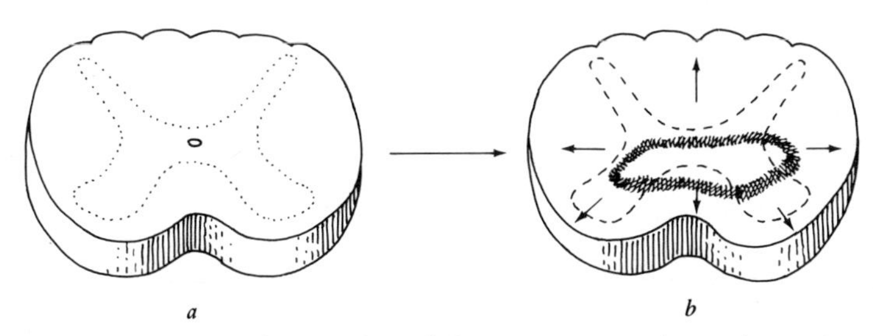

FIGURE 6-2. Cavitation in the spinal cord due to syringomyelia. *a* shows the normal appearance of the cord. *b* shows advanced cavitation, the arrows indicating the direction of future involvement.

A VICTORY FOR DR. SALK: POLIOMYELITIS

Poliomyelitis (*polis,* pale; *myelos,* marrow) is also known as polioencephalitis and infantile paralysis. Polio, as it is commonly called, is a viral disease that primarily affects the young, most cases occurring under age seven. The polio virus also has a special affinity for certain parts of the central nervous system, notably the anterior horns of the spinal cord and the motor cells of the medulla (bulbar polio), often spreading upward into the reticular formation and motor centers of the cerebrum.

During the acute phase of the disease, the anterior horn cells (or other motor cells) show progessive changes involving chromatolysis, with gradual destruction of the cells followed by phagocytosis of dead neurons (Fig. 6-3). Paralysis of the muscles innervated by the horn cells follows. Because the lower motor neurons are involved in anterior polio, the paralysis is of the flaccid type—usually in the lower limbs with atrophy of the affected members. When the polio virus invades the medulla, the disease is frequently fatal because of involvement of the respiratory centers.

Poliomyelitis is found throughout the world, occurring either epidemically or sporadically. Millions of individuals have been spared its ravages since the introduction of the Salk vaccine.

THE GREAT DISSIMULATOR: TABES DORSALIS

Tabes dorsalis (*tabere,* to waste away; *dorsalis,* dorsal) is one form of neurosyphilis, a disease now far less common as a result of the discovery of antibiotics. As its name implies, this disease has a special affinity for the tracts of fasciculus gracilis and fasciculus cuneatus (Fig. 6-4). The

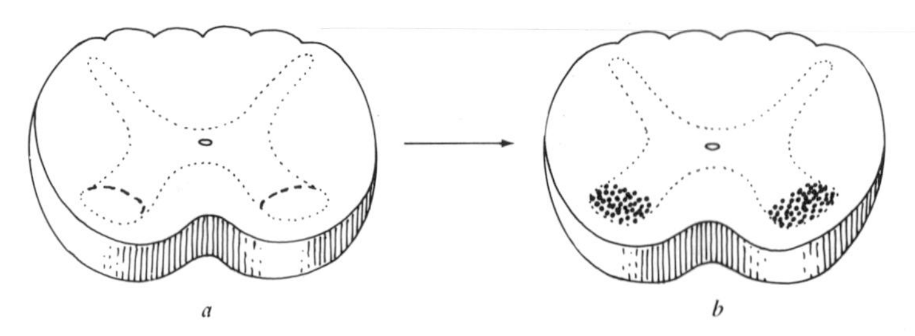

FIGURE 6-3. Destruction of the anterior horns by the poliomyelitis virus. *a* shows the spinal cord with the anterior horns demarcated. *b* shows how they are affected by the viral invasion.

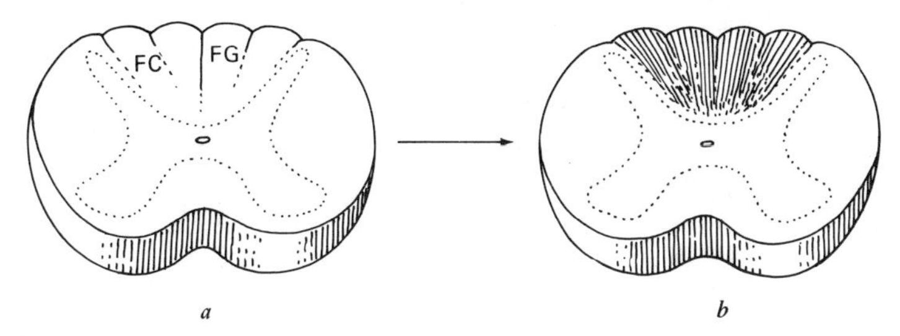

FIGURE 6-4. Destruction of the dorsal columns by *Treponema pallidum*. *a,* The normal cord with fasciculus gracilis (FG) and fasciculus cuneatus (FC); *b,* The same areas following destruction.

disease is caused by an invasion of the syphilis spirochete, *Treponema pallidum,* but does not appear in the spinal cord until from 5 to 15 years after the primary infection. Because the Wasserman test for syphilis is only positive in one-half the patients suffering from tabes, the syphilitic origin of the disease may not be apparent to the clinician on first examination. Because of its tendency to go underground and lie dormant for years producing a bewildering variety of symptoms when it surfaces, syphilis has been called the great dissimulator.

Since the dorsal sensory roots as well as the tracts of anterior and posterior funiculi are involved in tabes dorsalis, the patient experiences sensory disturbances and severe cramplike pains (tabetic crisis), which may be localized in the limbs and viscera as if they had originated in the receptors for dermatomes associated with the dorsal roots that are infected. The most prominent behavioral symptom is locomotor ataxia (*a,* not *tactos,* ordered), a disorder of coordination in which the individual slaps his feet down in a spasticlike gait while walking because of loss of muscle sensitivity mediated by the posterior funiculi (see Case IV).

A SAD FAREWELL: AMYOTROPHIC LATERAL SCLEROSIS (ALS)

The career of Lou Gehrig (1903–1941), one of America's most famous baseball players, was cut short by lateral sclerosis, a progressive degenerative disease of the lateral cortical spinal tracts and the anterior horn cells (Fig. 6-5). In a memorable occasion in Yankee Stadium, Gehrig bid a sad farewell to his teammates and fans.

Amyotrophic lateral sclerosis (ALS) is a disease of unknown origin that begins typically in the middle adult years and is characterized by muscular weakness, fatigue, obscure pains in the extremities, and tremor

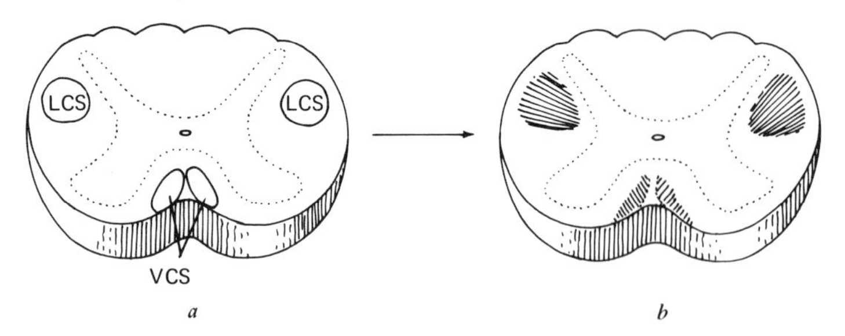

FIGURE 6-5. Lateral sclerosis. a, The normal cord with the lateral (LCS) and ventral (VCS) corticospinal tracts identified; *b*, The same areas following destruction by lateral sclerosis.

in the fine muscles of the fingers. As the disease progresses, muscular atrophy takes place as a consequence of the degeneration of motor nerves and the destruction of the ventral horn cells. The Babinski sign is present. Patients suffering from the disease typically show spreading involvement to the medulla and brainstem with disturbances of the muscles of the neck and face with consequent difficulty in swallowing, necessitating tube feeding in the final stages of the disease. Eventually the patient becomes paraplegic or quadraplegic. The prognosis for the disease is poor, with death occurring within three years.

A PUZZLING DISEASE: MULTIPLE SCLEROSIS (MS)

Multiple sclerosis (MS) is typically a chronic progressive degenerative disease involving not only the spinal cord but the brain as well. It is characterized by demyelinization of nerve fibers and proliferation of glia cells. The consequence is the appearance of plaques of sclerotic tissues that are scattered irregularly throughout the brain and spinal cord. The symptoms are unpredictable and variable with alternate remission and exacerbation of the disease. The classical triad of symptoms is nystagmus, intention tremor, and slow enunciation of speech with hesitation at the beginning of words or phrases. There may also appear tremors and weakness of the limbs, blurred vision, incontinence of the urinary bladder, and ataxia. Mental symptoms associated with the disease include periods of euphoria (exaggerated state of well-being), apathy, poor judgment, and severe mental deterioration in about one-fourth of the terminally ill patients.

In persons suffering from the chronic type of the disease, there may be long periods of remission that can last for several decades. The cause of the disease is not known. It is most common among young females in colder climates. Experimental treatments include the use of corticoids and vitamins. However, since there is no known cure, the disease eventually proves fatal.

7
THE SPECIAL SENSES

Étienne Bonnot de Condillac (1715–1780), a French philosopher, writing of the origins of mind, asks us to imagine a statue, perfect in form and endowed with all living organs except the senses. Such a creation, he argued, would be only a sham human being, a creature without consciousness or mind, a senseless and inert mass of flesh. In addition, he asks us to imagine further that the statue can be endowed with all of the senses, one after another. As each modality is added, the gradually awakening statue becomes more and more humanlike, conscious of its surroundings, curious, and responsive to stimulation. Alert at last, it can see, hear, feel, smell, and taste and is capable of the broad spectrum of reactions to its environment that make members of the animal kingdom so adaptable. But more importantly, Condillac reasons that his creation must also possess consciousness and a mind. For, true empiricist (*empeirikos,* experience) that he was, he believed that the elements of mind are sensory experiences. Ideas are nothing more than the residuals of sensation; and in association with each other, they account for all the complexities of mental life.

We will never know if by some magic Condillac might have endowed a statue with life, and perhaps fallen in love with his own creation as did Pygmalion, but we can agree that a senseless creature, unresponsive to its environment and unable to relate one experience to another, would be little more than a vegetable—a term often used to characterize the comatose or moribund. And so, before we describe the brain and its associated cranial nerves, we need to know something of the special senses that make up a large portion of the input system for the brain and on which we depend for most of the information about our environment.

VISION: IN THE EYE OF THE BEHOLDER

The eye is a hollow chamber, slightly bulged in front (Fig. 7-1) and filled with fluidlike substances known as the aqueous (*aqua,* water) and vitreous (*vitreus,* glass) humors. The aqueous humor is secreted constantly by the ciliary body and contains metabolites for the nutrition of the cornea. The aqueous humor is drained through an opening called the canal of Schlemm where it enters the venous system (see Case VI.) The vitreous humor is a transparent permanent gel that helps lend form and stability to the eye. As the cross-sectional diagram shows, the eye consists of three layers, the sclera (*sklera,* hard), a tough outer coat that gives form to the organ. In front of the eye the sclera is modified into the cornea (*cornea,* shield), a transparent shield full of pain receptors that also begins the process of focusing or refracting light.

The choroid coat (*choroeides,* chorion, placental tissue) is a highly vascular and darkly pigmented layer that nourishes the receptor cells

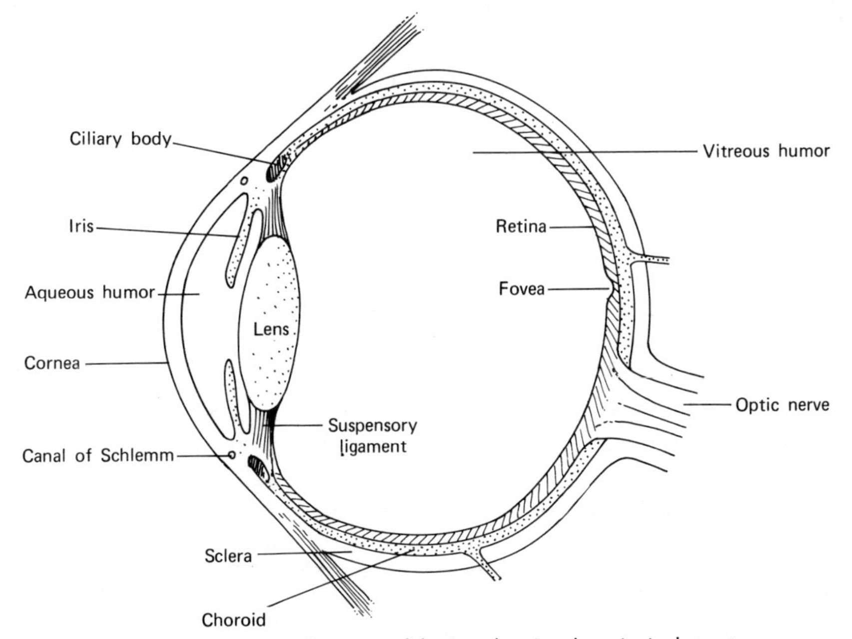

FIGURE 7-1. A saggital section of the eye showing the principal structures.

with nutritive products from the blood and also absorbs light that would otherwise scatter throughout the eye and blur vision. The retina (*rete,* net) is a network of receptor cells and their associated connecting neurons whose function we shall examine in detail later.

The iris (*iris,* rainbow), a pigmented structure whose color and form has often inspired romantic feelings, is a contractile membrane containing both circular and radial fibers whose function is to control the amount of light entering the eye. The pupil of the eye is an opening in the center of the iris through which light passes. Under conditions of twilight (or during strong emotion), the sympathetic nervous system stimulates the dilator muscle of the iris and the pupil dilates. In bright daylight (or during periods of relative calm) the pupil constricts when parasympathetic stimulation causes its contraction. Italian ladies of bygone days used to put drops of extract of belladonna (*bella,* beautiful; *donna,* lady), a common plant, in their eyes to dilate the pupils and so enhance their appearance.

The lens is a clear, oblong structure suspended behind the pupil. Its function is to complete the focusing or refractive process initiated by the cornea. For distant objects the lens is flattened; for near objects, it becomes more spheroid. This process of accommodation is controlled by the ciliary body, a collection of muscles attached to the lens by the suspensory ligaments. The muscles of the ciliary body are under reflex control of optic centers in the brain (see Fig. 7-3). One of the early signs that the individual is growing old is the loss of the ability to accommodate—often first observed when the arm seems too short to

CASE VI

IRIDECTOMY FOR THE RELIEF
OF GLAUCOMA

Glaucoma is a disease of the eye characterized by excessive pressure in the anterior chamber. This may be caused either by a defect in the canal of Schlemm, a small opening on the edge of the iris that allows excessive aqueous humor to flow to the surface, or to excessive production of aqueous humor exceeding the drainage capacity of the canal. In either case the increasing pressure endangers the retina, with blindness eventually resulting. Primarily a disorder of the middle years, glaucoma is a leading cause of blindness.

G.E., a fifty-one-year-old woman, was admitted to the medical center for peripheral iridectomy to relieve inadequate aqueous drainage caused by a narrowing of the canal of Schlemm, a venous return system for the aqueous humor. Following surgical preparation and the administration of a local anesthetic, a small incision (see *1*) was made on the surface of the cornea at the edge of the iris at about 11 o'clock, frontal view. The incision was made sufficiently deep to allow extrusion of a knuckle of the iris when pressure was applied *(2)*. The knuckle was then cut *(3)* Depressing the posterior lip of the iris allowed aqueous humor to flow outward from the anterior chamber *(4)*. The anterior lip of the iris was repositioned by depressing it, providing for a larger channel for drainage *(5)*.The frontal appearance of the eye after surgery is shown in *(6)*.

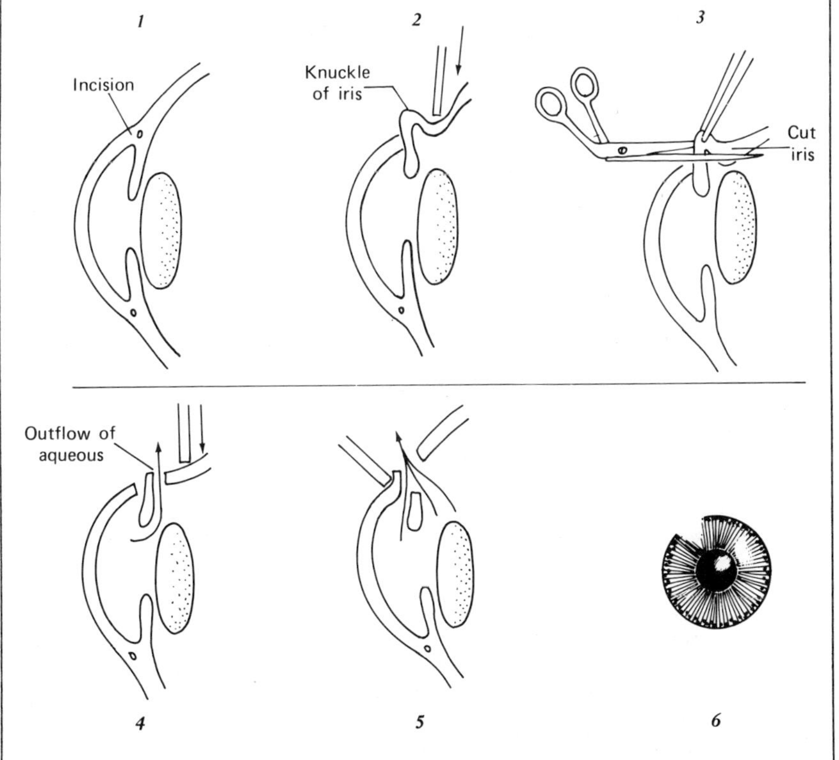

Following the operation, G.E. showed no further symptoms of increased intraorbital pressure, and the danger of retinal damage was eliminated.

bring the telephone directory into focus. Said to have been invented by the versatile Benjamin Franklin, bifocal lenses, with a lower half for near vision and the upper half for far, are a very satisfactory solution to this particular problem of aging.

The retina is a complex membrane comprised of three layers of cells (Fig. 7-2). The innermost layer nearest the choroid coat is made up of rods and cones, the receptor cells for vision. These are connected to bipolar cells which, in turn, synapse with ganglion cells or the optic neurons that form the optic tract. Cones are typically connected to individual bipolar cells, while a number of rods may be served by a single bipolar cell. The diagram also shows horizontal cells that connect the receptors and amacrine (*ama,* a vase; *crine,* tentacles) cells that interconnect the bipolar layer. This arrangement makes possible interaction

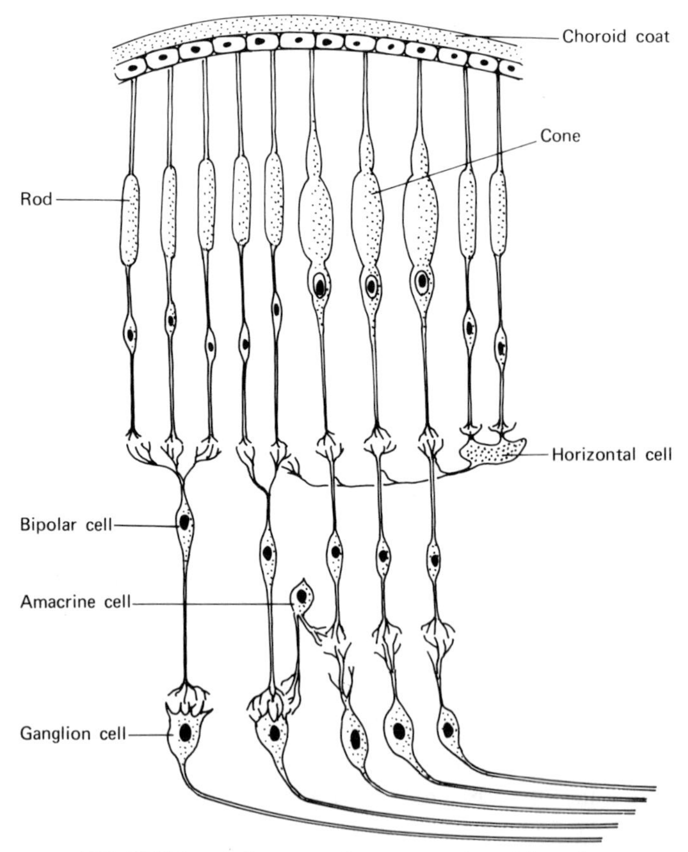

FIGURE 7-2. A diagrammatic representation of the retina.

among the retinal elements, a process we shall have more to say about later.

IN BLACK-AND-WHITE AND COLOR

From a mass of evidence assembled over the past century, we know that rods mediate twilight or brightness vision, and cones, daylight or color vision. Rods, therefore, are the receptors active in dim light or at night and are only capable of poor detail resolution. Cones are our daylight receptors and have a high degree of ability to discern detail, a fact that is related to the one-to-one connections between cones and bipolar cells. It is the acuity of cone vision that the eye specialist measures with his graded series of letter charts.

The fovea (*fovea,* pit), a small depression at the rear center of the retina, contains only cones and is the center of clearest daylight vision. For this reason we turn our eyes to focus images on the fovea when we wish to attend to a visual stimulus. At night the fovea is insensitive, and this is why sentries are told to look a little indirectly at suspicious sources of sound. It is instructive to try this experiment on a clear night with a faint star that will brighten significantly if its image is moved a few degrees off center. However, the rods that are most numerous around the periphery of the retina, even though incapable of responding with sensations of color or high acuity, are endowed with great sensitivity for seeing in dim light. Everyone has experienced dark adaptation—the increasing ability to see objects more clearly as time passes after entering a darkened room, such as a theater. Most dark adaptation takes place within one-half hour after entering darkness during which time the rods become over 100,000 times more sensitive to light. Indeed, some physiologists have calculated that under ideal conditions of dark adaptation a single rod might be sensitive to one photon of light, the smallest measurable quantity.

The action of the rods depends upon the presence of a photopigment called rhodopsin (*rhodo,* rose; *opsin,* appearance), or visual purple, which is found around the rods near the choroid coat. Light bleaches out rhodopsin, much as it does a photographic film. But unlike photographic films, the eye is capable of regenerating photochemicals. Consequently, in the presence of darkness or dim light, visual purple regenerates. Because vitamin A is necessary to the regenerative process, its absence may lead to night blindness—a discovery first made by ancient Egyptian physicians who prescribed raw liver to correct the condition. Happily, we now know more palatable sources of this vitamin.

Before considering the visual pathways to the brain, we might note that each eye has an optic disc, an area where the optic neurons gather as they leave the retina. This area is called the blind spot, because the crowding that results from the collecting nerve fibers leaves no room for sense cells. If you close your left eye and hold a pencil about a foot out in front of the right eye—a little to the right of center—and slowly move it around, you will find a spot where the eraser disappears. Normally, we do not notice our blind spots because of overlapping visual fields and the tendency of the brain to fill in gaps in perception.

FROM EYE TO BRAIN: X MARKS THE SPOT

The optic neurons leaving the eye from the two optic nerves or tracts meet under the brain at the optic chiasma (*chiasein,* to mark with an X). There a strange crossing takes place. Half of the neurons—those from the temporal half of each retina—stay on the same side; and the other half—those originating from the nasal side of each retina—cross (Fig. 7-3). Because of this arrangment, the right side of the brain receives fibers from both the right and left retina. For this reason a destruction of one optic tract between the chiasma and its termination in the brain results in a hemianopsia (*hemi,* half; *ops,* eye) or half vision, or if you prefer, half blindness. If the right tract is interrupted, the right one-half of each retina will be affected. Similarly, if the left tract is interrupted, the left half of each retina is cut off from the brain.

As Figure 7-3 indicates, the optic fibers synapse in the lateral geniculate nuclei (*geniculum,* knee: from the bent shape of the nucleus). From there, most fibers go directly to the occipital cortex, but some have connections to the superior colliculi, the oculomotor, and pretectal nuclei, which are concerned with adjusting the diameter of the pupil to varying degrees of brightness and the lens for accommodation for near and far vision. The accommodation process is accomplished by the pretectal nuclei, which have fibers that connect with the ciliary ganglia located near the eye. In turn, the eye eventually sends fibers to the ciliary muscle.

WHAT THE EYE TELLS THE BRAIN

A rose has color. It has a distinct shape. It also stands out as a figure from the background of foliage. A bee feasting on nectar suddenly flies away home. How much of all this information does the eye supply and how

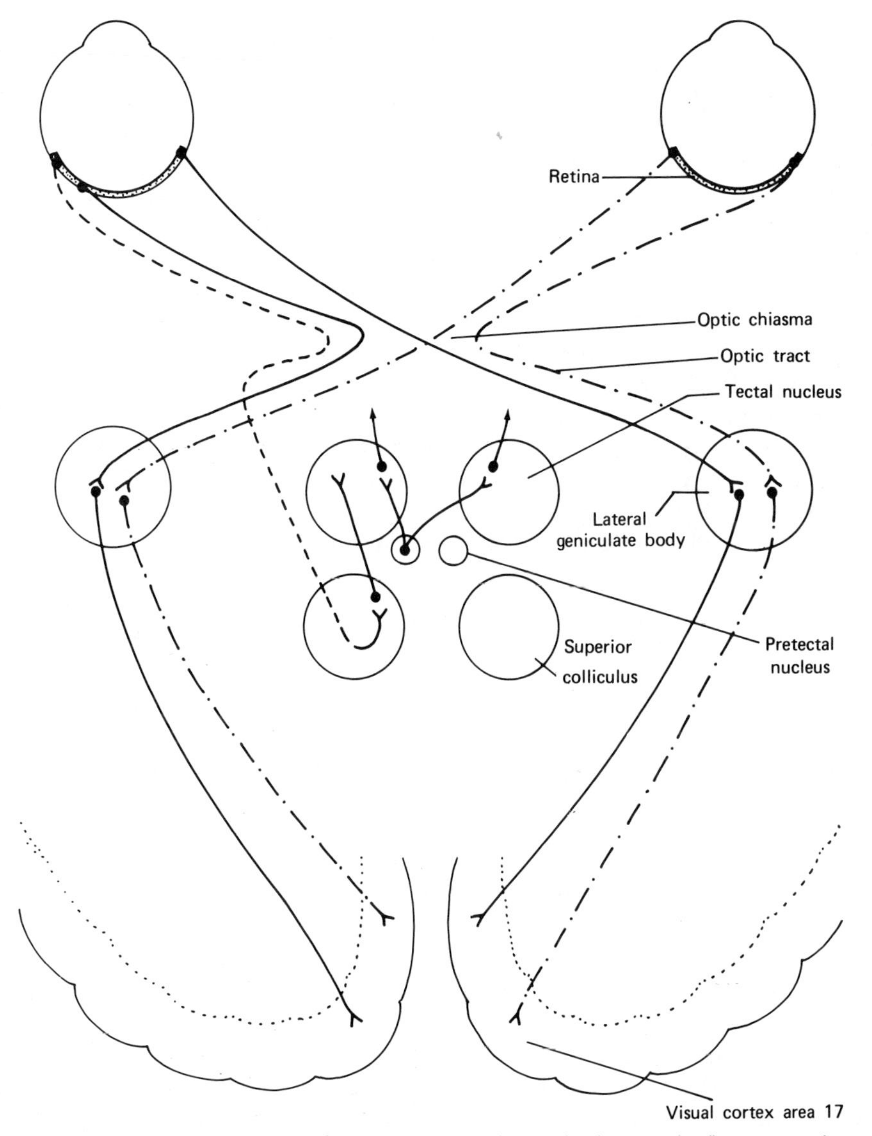

FIGURE 7-3. The optic chiasma and subcortical nuclei involved in visual reflexes. Note that nasal fibers cross while temporal fibers do not.

much depends on the brain's learned interpretation of neural impulses that arrive from the two retinas? In recent years knowledge has been amassing rapidly to explain how information about objects is transduced (*transduce,* change from one form of energy to another) from the receptors, coded into nervous impulses, and sent to the cortex. Microelectrode studies of the cones (Experiment VI) have proved that there are three basic types of cones that are especially sensitive to certain wavelengths of light and that, when transduced into nervous impulses, give rise to the experiences of red, green, and blue. Mixtures of these wavelengths arouse other color experiences, including white, which Newton's famous prism proved is a mixture of all wavelengths.

We also know from studies of the lateral geniculate nucleus that coding goes on in the retina so that color impulses are transmitted in pairs: red-green and blue-yellow, which are complements or antagonists, because when these are mixed in proper proportions (using lights, not pigments) they yield grays. Some neurons appear to be blue excitatory (+B) and yellow inhibitory (−Y). Some are opposites: +Y and −B. Similarly, there are +R and −G, and +G and −R codes. We are not certain why the retina codes its responses to the primary wavelengths in this manner, but it appears to be consistent with a great deal of inhibition and facilitation that characterizes retinal activity. For example, there is also a retinal code for on-off. Some optic neurons fire on being stimulated with light. Others fire when the light goes off. And still others maintain a steady firing so long as the light remains on. We also know that excitation of one set of retinal cells tends to suppress activity in surrounding cells. These opponent processes may insure that figures in the foreground stand out in seeing while background information is suppressed or not transmitted. Moreover, the retina is capable of spatial summation as well as inhibition. The presence of the horizontal cells and of amacrine fibers makes possible additive effects so that two weak stimuli, neither of which would be transmitted singly, can trigger a message by summating.

Research on lower forms has also shown that the occipital cortex receives codes that tell it whether lines are horizontal, tilted, or vertical and whether the stimulus stands out as an edge or figure or is part of the background. There are also codes to tell whether an object is still or moving. In the frog—a creature noted for its ability to snare flies on the wing with a long sticky tongue—some cells specialize in informing the frog's brain about small black objects moving across the field of view. These same cells ignore small black objects that do not move.

The eye therefore is not a passive photochemical transducer but a

EXPERIMENT VI

MacNICHOL AND WALD DISCOVER THE RECEPTORS FOR COLOR

In 1863 the great German physicist and physiologist, Hermann L.F. von Helmholtz, who carried out the landmark experiment of measuring the speed of the nervous impulse in 1851 (see Experiment I), announced his trichromatic or three-color theory of vision. He reasoned that there are three types of cones in the retina—red, green, and blue—each maximally sensitive to certain wavelengths—red to the long waves, green to the medium, and blue to the short. Helmholtz chose these three colors because they are primaries. All other colors, including white, can be obtained by mixing two or more of the primaries in proper proportions.

Helmholtz's reasoning was logical and fit color experience very well, but hard proof for the essential validity of the theory had to wait about a century for the research of E.F. MacNichol, Jr., of Johns Hopkins, and George Wald of Harvard. MacNichol devised a microspectrophotometer, or in simple terms, a device for stimulating a single cone with a microscopic dot of light while measuring its reaction to the stimulus. He found three basic types of cones, each maximally sensitive to wavelengths corresponding to what we experience as red, green, and blue, just as Helmholtz had predicted.

Wald's Nobel prizewinning research on the photochemistry of the retina led to the discovery of three types of iodopsin, the photochemical responsible for transduction in cones as rhodopsin is the transducer for rods. The three basic types are: erythrolabe (red sensitive), cynolabe (blue sensitive), and chlorolabe (green sensitive).

The researches of MacNichol, Wald, and their associates are brilliant examples of the application of modern technology to unraveling the mysteries of the sense organs. They are also eloquent testimony to the value of a good theory in directing research into productive channels.

complex extension of the brain that furnishes cortical centers with considerable information in the form of codings about color, shape, movement, and possibly spatial depth. Far from being a simple receptor, the eye is an information-processing center of considerable sophistication.

TO THE CORTEX: BRIEFLY NOTED

Neurons from the optic tract terminate in a small area of the occipital cortex known as the primary visual area or area 17. There they synapse with other neurons that spread over a wide region of the occipital lobe of the cortex and have potential connections with all other parts of the brain as well. We shall be considering these centers further in Chapter 16, which is concerned with the functions of the cerebral cortex.

AUDITION: AN EARFUL OF BONES AND SHELLS

The human ear is specialized for three functions: the collection of sound waves, their amplification, and their transduction into nervous impulses. The collecting system is the outer ear; amplifying is the function of the middle ear; and transducing takes place in the inner ear (Fig. 7-4).

Our outer ears consist of three parts: the pinna (*pinna,* wing) or auricle; the external auditory canal; and the tympanic (*tympanos,* drum) membrane or ear drum. In man the pinna serves little purpose, except perhaps to support spectacles or earrings. In animals that have control over the directionality of the pinna it is undoubtedly a significant aid for localizing sounds. Watch the ears of a deer, if you are fortunate enough to see one, or even those of a domestic dog. The external auditory canal funnels sounds to the tympanic membrane, which is set into sympathetic vibration or resonance by those sounds. The effect in the ear is similar to that experienced when holding a metal wastebasket near a loudspeaker on a high fidelity amplifier. The basket will be felt to vibrate or resonate with the speaker. The highly flexible and sensitive tympanic membrane can respond to incredibly small disturbances of the surrounding air, as minute, in fact, as one-billionth of a centimeter. If the ear were more sensitive, we would be unable to hear distinctly, for we would be confused by the Brownian movement of air particles and by the sounds of blood rushing through the auditory capillaries.

In the middle ear are three tiny bones or ossicles (Fig. 7-4). Attached

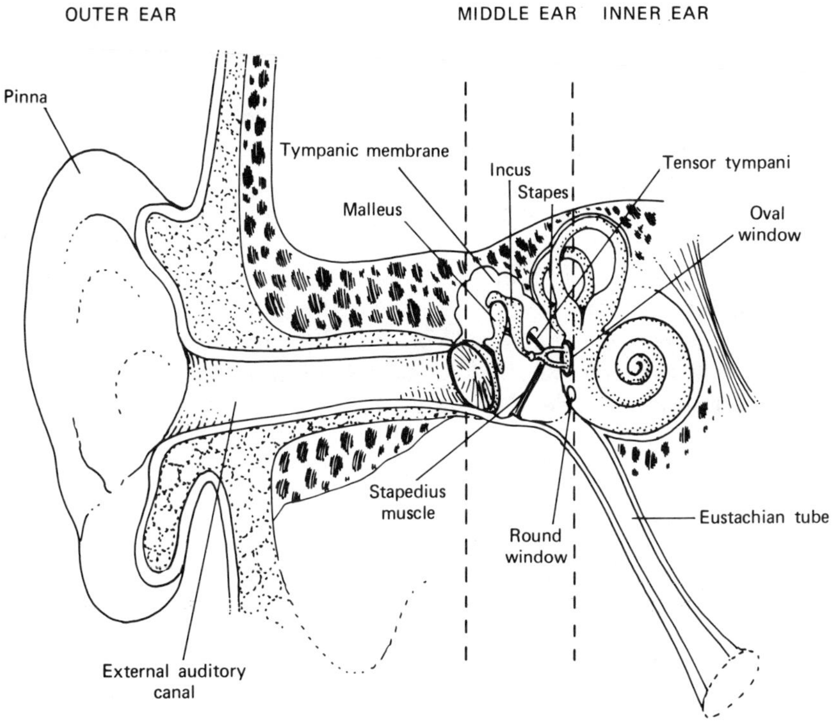

FIGURE 7-4. Gross anatomy of the ear showing the principal structures of the outer, middle, and inner ear.

to an opening in the inner ear called the oval window, is the stapes (*stapes,* stirrup), shaped like the object for which it is named. The stapes articulates with the incus (*incus,* anvil), which, in turn, connects with the malleus (*malleus,* hammer). This miniature set of blacksmith's equipment is arranged in a v-shape and is therefore capable of acting as a bent lever for amplifying sounds.

There are two muscles that control the ossicles. One, the tensor tympani, is attached to the malleus in such a way as to put tension on the tympanic membrane and so dampen excessive vibrations. The stapedius muscle attaches to the stapes to dampen any excessive contractions of this ossicle. Another protective mechanism for the tympanic membrane is the Eustachian tube (named after the anatomist, Eustachio), which connects with the back of the mouth. We often feel its valve "popping" to allow pressure equalization across the eardrum, otherwise very loud sounds (or changes in atmospheric pressures as when going up a mountain) could rupture the membrane. This possibility, incidentally, is what cannoneers avoid by opening the mouth and shouting when siege artillery is being discharged.

The mechanisms of the inner ear are contained in the cochlea (*kochlias,* snail or shell fish), a coiled bony structure resembling a snail's shell. Inside are three fluid-filled ducts or canals: the vestibular, the cochlear, and the tympanic canal. The arrangement of the canals is shown in Figure 7-4. Note that the tympanic canal ends at the round window, a membranous opening in the bony cochlea. Because the fluid in the cochlea is relatively incompressible, the bulging in and out of the round window is in synchrony with the same movements at the oval window. This arrangement allows sound waves to travel through the fluids of the inner ear and to stimulate the hair cells of the organ of Corti.

CASE VII

STAPEDECTOMY FOR THE RELIEF OF CONDUCTION DEAFNESS

The two basic types of deafness are (1) conduction deafness caused by defects in the conduction mechanism of the middle ear, and (2) nerve deafness resulting from injury or disease to the auditory nerve. As yet there is no cure for nerve deafness, but a surgical procedure has been developed to alleviate many cases of conduction deafness.

O.S., a thirty-six-year-old salesman, complained of greatly decreased hearing in his right ear. The condition was not painful but had become socially distressing and threatened his livelihood. Examination of the ear revealed a tightened or retracted tympanic membrane caused by a history of childhood infections in the throat and extending up the Eustachian tube, which had necessitated numerous punctures of the tympanic membrane. Moreover, the malleus was not appreciably movable, suggesting that the ossicles might be fixed owing to degeneration.

The middle ear was opened under local anesthesia and the following procedure carried out under a binocular microscope. (1) The incus and stapes were separated since that joint was fixed (see point *a* in sketch *1*). (2) The posterior portion of the stapes was separated as shown between the dotted lines, with the midportion removed (*a* and *b*). (3) The anterior portion of the stapes was fractured at *c* leaving a section partially fixed to the tendon of the stapedius muscle. The other portion was removed. (4) The footplate of the stapes was removed from the oval window at *d*. (5) A graft from the perichondrium—a highly vascular covering of the cartilage of the small flap of the pinna that shields the opening to the external canal—was sewn into place in the new oval window (see sketch *2*). (6) The remaining portion of the stapes was reoriented to fit into the center of the graft. (7) The head of the stapes, previously separated from the incus, point *a*, sketch *1*, was relocated under the incus in a normal anatomical relationship, sketch *2*, thus reestablishing the conduction chain.

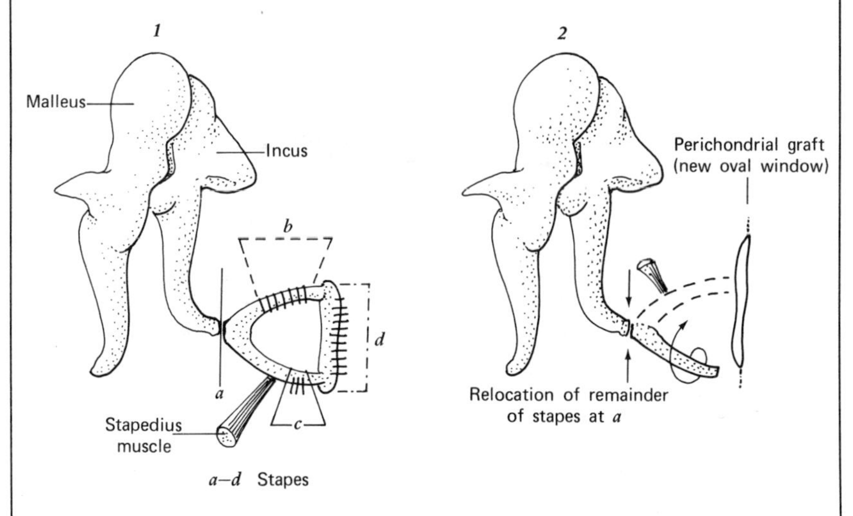

At this point in the operation, the patient's hearing improved dramatically, the surgeon no longer having to raise his voice for the patient to hear clearly.

Following the operation, O.S. suffered no undesirable side effects, such as dizziness or ringing in the ears, and enjoyed a pronounced increase in auditory acuity.

The organ of Corti (named after its discoverer) is the organ of hearing (Fig. 7-5). Essentially, it consists of hair cells located along the basilar membrane. There are approximately 25,000 of these cells in the outer layer and approximately 3500 in the inner layer. Resting on top of the hair cells is the tectorial membrane (*tectum,* roof). The auditory nerve fibers lead off from the hair cells to the spiral ganglion on their way to the brain.

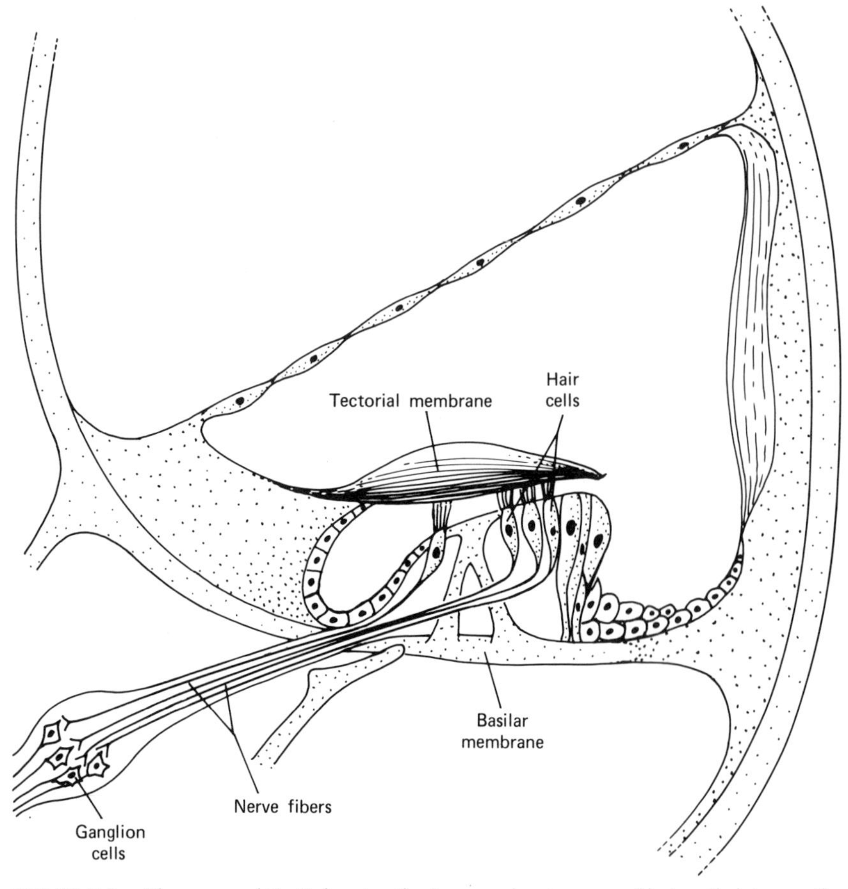

FIGURE 7-5. The organ of Corti showing the inner and outer rows of hair cells lying on the basilar membrane.

HOW IT ALL WORKS: RIPPLES IN THE POND

When sounds enter the inner ear by way of the oval window, waves are generated in the fluids of the vestibular canal much as waves are generated in a pond of water by a thrown stone. As these travel along toward the end of the canal, they push upward against the basilar membrane, bending or shearing the hair cells against the tectorial membrane. The bending of the cells generates a potential change, which summates or is added to the resting potential of the basilar membrane to trigger nervous impulses.

The mechanism of coding for pitch, or the quality of highness and lowness by means of which we characterize sounds, has been a difficult puzzle for generations of neurophysiologists beginning with Helmholtz. He suggested that the basilar membrane, which is shaped like a harp or piano sounding board, resonates at different places depending on the pitch of the sound entering the ear. High-pitched sounds stimulate the shorter fibers of the membrane, low pitches stimulate the longer end. Because of his assumption about the resonating functions of the membrane, Helmholtz's theory has been called the piano or place theory. According to a place theory loudness, the other most important dimension of auditory experience, is the result of the number of cells firing at any given place.

In 1866, following the invention of the telephone, William Rutherford, a physicist, argued that the ear responds to incoming frequencies like a telephone receiver responds to the voice on the other end. In other words, the tympanic membrane and basilar membrane vibrate like the diaphragm in a telephone receiver mimicking external sounds.

Rutherford's clever analogy was partially confirmed in the 1930s when scientists at Harvard discovered that for the lower end of the pitch range, from 20–4000 Hz (Hz = Hertz, or cycles per second) the auditory nerve does, in fact, follow the frequency of the stimulating sound. However, for higher pitches 4000 Hz and up, to the limit of hearing or about 20,000 Hz, the place stimulated by the traveling wave in the cochlear fluids appears to be the critical factor.

What we have today is a combination theory, a place theory and a modified frequency theory known as the volley theory. The latter attempts to account for the fact that neurons of the auditory nerve can follow sounds of frequencies up to 4000 Hz; since, when firing as fast as possible, a single neuron can only fire up to 1000 times per second. The volley theory assumes that there are squads of neurons that fire in rotating volleys, just as squads of soldiers in a platoon can fire in sequential volleys. This kind of firing would be faster than if all neurons (or soldiers) fired simultaneously, having to rest during the refractory period (or during reloading) and then fire again. According to this theory loudness can be accounted for by the number of neurons firing in each volley.

FROM EAR TO BRAIN

Figure 7-6 shows the auditory path from the cochlea to the brain in simplified form. The fibers of this branch of the eighth nerve, one of the cranial nerves, first travel to the cochlear nucleus synapse, before

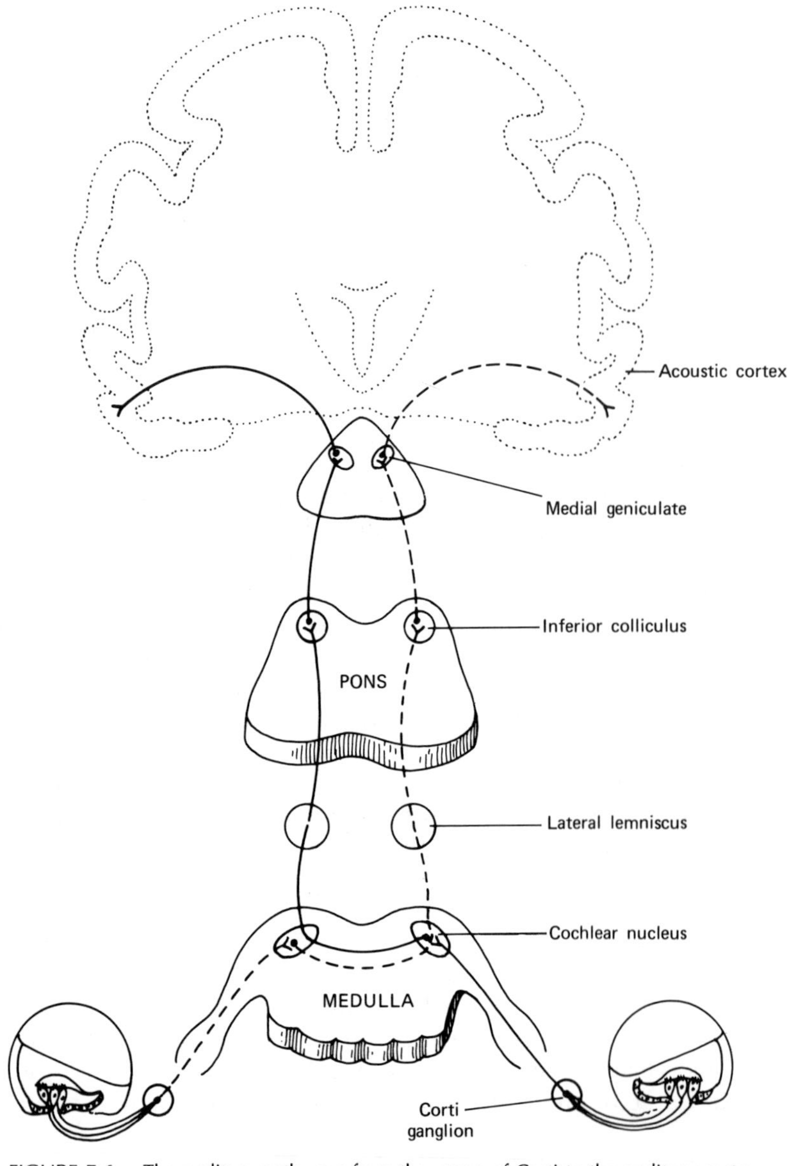

FIGURE 7-6. The auditory pathways from the organ of Corti to the auditory cortex.

crossing to synapse again in the inferior colliculus on the opposite side. From there they proceed to the medial geniculate bodies and finally to the cortex. As a result of this arrangement, the auditory system, like the visual system, provides for bilateral representation of both ears on each side of the brain. We shall be considering the further connections of the auditory system in Chapter 16.

A SENSE OF BALANCE: THE VESTIBULAR SYSTEM AND LABYRINTHINE MECHANISMS

We marvel at the grand finale of the ice skater who spins around so rapidly he or she is a blur, then suddenly stops, and without the least sign of distress or incoordination, smilingly bows to the audience. We know that if we attempted to emulate such a performance, we would become dizzy, stumble or fall down, and might disgrace ourselves by becoming nauseated. In fact, many people experience symptoms as severe as these with far less provocation, for example, when traveling on a ship or riding in an automobile. Disturbances of balance and equilibrium are caused by strong stimulation of the vestibular mechanisms, which are closely associated with the inner ear. The receptors of this system are located in the semicircular canals and in two membranous sacs, the utricle (*uter,* sac or bag) and the saccule (*saccus,* a sac). These are shown in diagrammatic form in Figure 7-7. We shall describe the structure and function of each of them separately.

THE SEMICIRCULAR CANALS

The semicircular canals are small membranous tubes inside a bony labyrinth. The tubes are filled with a fluid called the endolymph (*endo,* with; *lympha,* water) and are cushioned from the surrounding bony labyrinth by a similar fluid, the perilymph (*peri,* around). The three ducts are related to each other in such a manner that each occupies a different plane in space. For this reason, acceleration or deceleration of the head in any direction will stimulate the fluids in one of the canals.

As Figure 7-8 shows, the crista (*crista,* crest or ridge), which is similar in shape to the organ of Corti, contains hair cells that are formed into a crest embedded in a gelatinous mass, named the cupula (*cupula,* dome) for its domelike appearance. Because of inertia, the fluid in the appro-

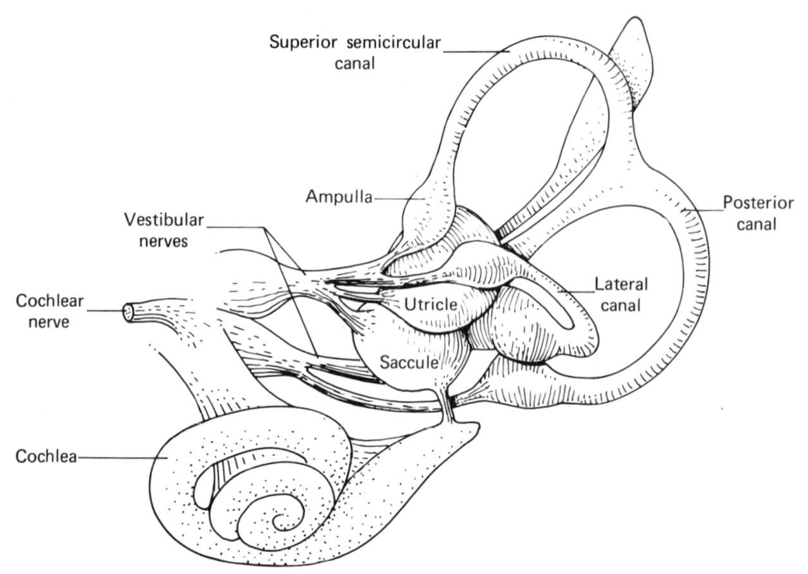

FIGURE 7-7. The labyrinthine organs showing the relationship of the semicircular canals, the utricle, and saccule to each other and to the cochlea. The view is of the left side.

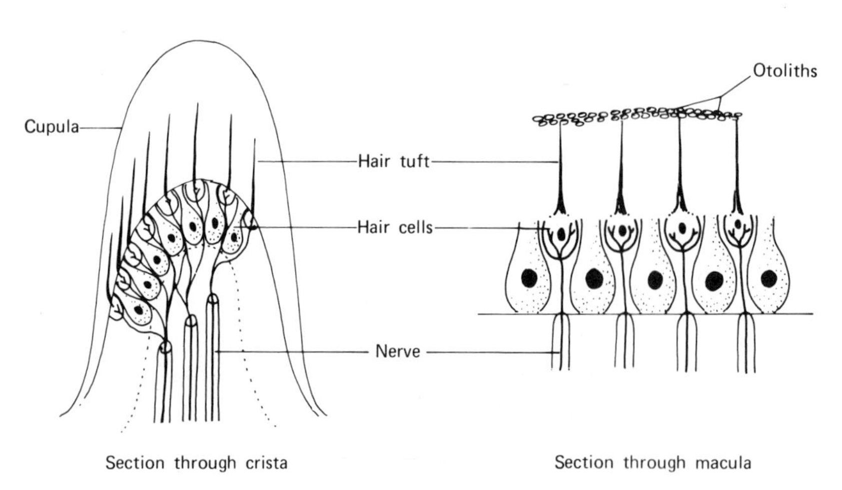

FIGURE 7-8. Diagrammatic view of the crista and macula.

priate canal tends to remain where it was when the head is moved, thus exerting pressure on the cupula. When the head stops moving, the fluid tries to continue moving, again exerting pressure on the cupula but this time in an opposite direction. With continuous movement of the head, stimulation stops. Therefore, it is change in angular acceleration that stimulates, that is, a change in the speed of rotation of the head. For example, if the head is suddenly turned toward the right in a horizontal plane, the fluid in the horizontal canal shows inertia, lagging behind the movement of the canal. The fluid, therefore, pushes against the receptor. Rotation in other planes would similarly stimulate the receptors in other canals.

When pressure is exerted on the cupula from a change in angular acceleration, the hair cells transduce this mechanical energy into nervous impulses that are transmitted over the vestibular branch of the eighth nerve. We shall trace this pathway to the brain after describing the functions of the utricle and saccule.

UTRICLE AND SACCULE: PURSES FILLED WITH PRECIOUS CRYSTALS

The receptor cells of the utricle and saccule are contained in the macula (*macula,* a spot or stain) so named because the hair cells are embedded in a gelatinous mass containing otoliths (*oto,* ear; *lith,* stone) of calcium carbonate crystals that give it a spotty appearance. The macula of the utricle is horizontal when the head is in its normal position, and the macula of the saccule is vertical in normal head position. When the head changes orientation, the heavy gelatinous mass above the hair cells exerts a strong pull, changing the pattern of nervous impulses and thereby making available information about head orientation.

It is also believed that the macula of the utricle may be sensitive to linear acceleration, that is, to start-stop movements of the head.

The nerve fibers from the utricle and saccule join those from the semicircular canals to form the vestibular nerve, which terminates in the vestibular nucleus of the brainstem. From there, impulses that function in righting reflexes are relayed to the spinal cord over the vestibulospinal tract and to nuclei that supply motor impulses to the extrinsic eye muscles. Some fibers also connect with visceral nuclei, which accounts for the sense of nausea sometimes experienced if the vestibular mechanisms are strongly stimulated. The connection to centers for the control of eye muscles make possible the regulation of eye movements to compensate for head movements. Impulses that function in righting reflexes are observed most dramatically in nystagmus (*nystagmos,* nodding), an involuntary oscillation of the eyeballs following head rotation. Nystagmus is an attempt to fixate the eyes during rotation and will disappear if rotation continues, but then reappears briefly after rotation ceases.

Nystagmus can be utilized as a clinical test for the integrity of the semicircular canals and vestibular system. There are two methods for accomplishing this. In the first, the individual is seated in a chair, similar to a barber's chair, that can be rotated rapidly with the head angulated in various positions during successive tests of each of the canals. When the chair is stopped, nystagmus occurs with a slow movement of the eyes in the direction of rotation and a fast component in the opposite direction. During nystagmus, which lasts for 15–20 seconds, the individual feels as if he is still rotating. Failure of nystagmus to occur in any plane indicates a disorder in that canal. This technique is limited in that both sides of the head are involved simultaneously.

The irrigation method for a separate test of each side consists of placing cold water in the external canal of one ear. This causes convection currents in the canal, which stimulate in turn the receptors bringing on nystagmus. By tilting the head appropriately, each canal can be examined separately for integrity of function.

The phylogenetically ancient sense of balance and equilibrium is probably more important in birds, fishes, and other forms who are likely to find themselves flying or swimming under conditions of poor visibility or in unstable air or water. In man and the higher vertebrates whose feet are on the ground, the sense functions mainly to assist vision and kinesthesis.

THE CHEMICAL SENSES

Taste and smell are known as the chemical senses because the stimuli for their receptors are substances in solution, in the case of taste, and in gaseous form, in the case of smell. The two senses are also related in that much of what we think we taste we actually smell, as is made abundantly evident during a siege with a bad cold. To complicate matters still further, the quality of things tasted is made more complex by contributions from the other senses in the oral cavity, touch, temperature, and kinesthesis—even pain for those who fancy powerful condiments in their food. However, the two senses are separate modalities, with specialized receptors and cortical connections, which we shall now summarize.

GUSTATION: A MATTER OF TASTE

The receptors for taste are located mostly in the mucous membrane of the tongue but a few are found in the membranes of the pharynx, palate, and larynx. They are commonly referred to as taste buds from their appearance under the microscope (Fig. 7-9). Each taste bud is equipped with hairlike projections called microvilli that project upward through the taste pore and that, when stimulated by substances in solution, are transduced into nervous impulses. There is reason to believe there may be different types of buds mediating different taste qualities, since the tongue is differentially sensitive to sweet, salt, sour, and bitter. Sweet and salt sensations arise primarily from the tip of the tongue, sour along the edges, and bitter near the base of the tongue. This is why the bitter taste of quinine is at its worst after it has been swallowed. However, microelectrode recordings from the taste nerves do not support a simple, discrete type of coding. Instead, there are complicated patterns of firing, with some neurons responding maximally to certain solutions representative of sweet, salt, sour, or bitter, but also showing some responsiveness to all other solutions. Evidently the brain in some not yet completely understood manner can decode the complex patterns into what we experience as discrete tastes.

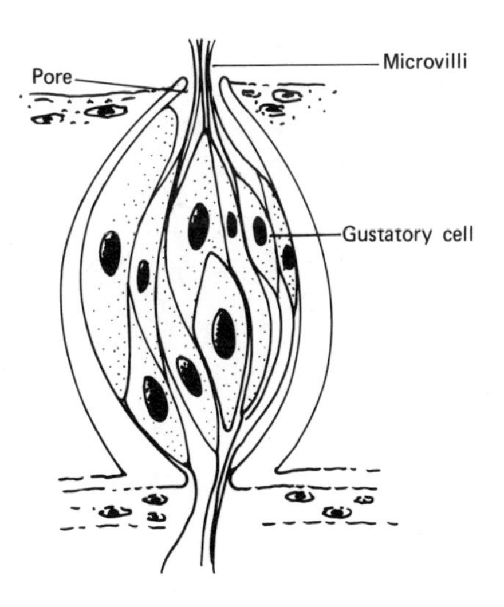

FIGURE 7-9. A taste bud.

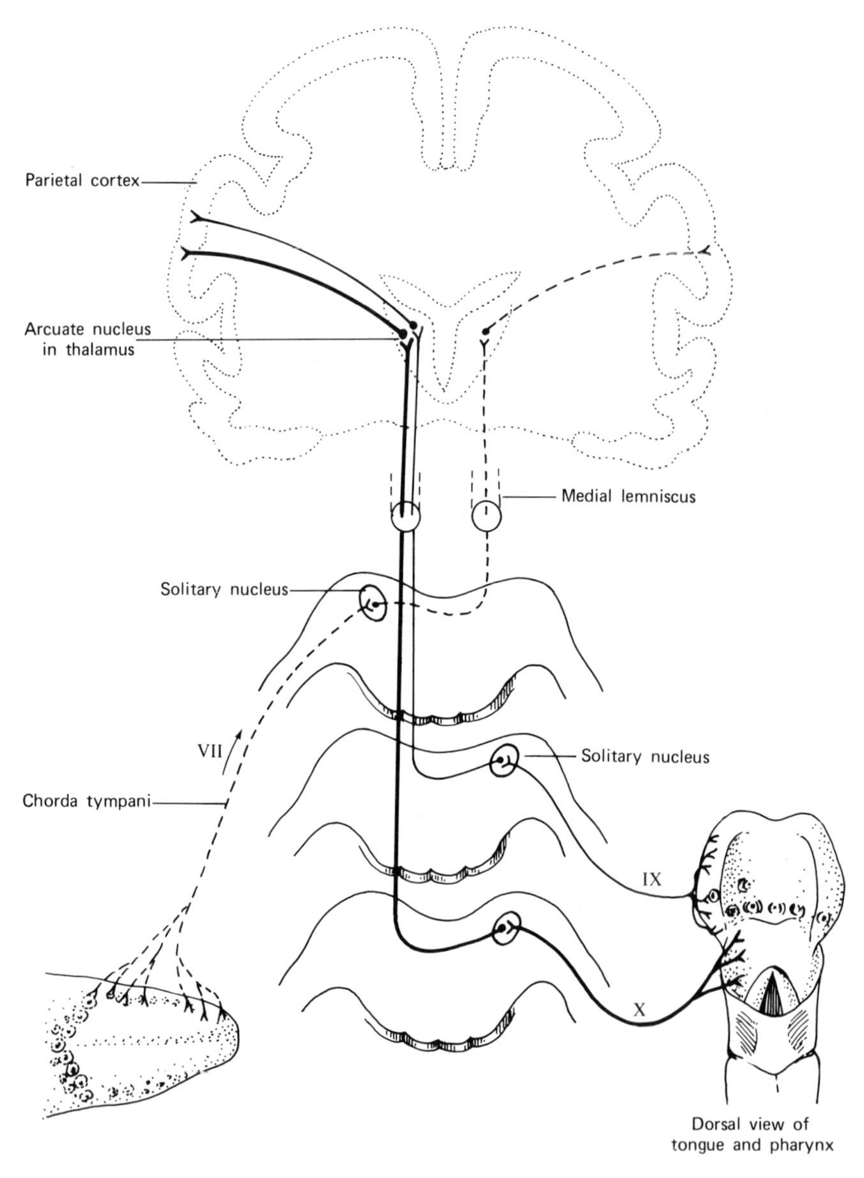

FIGURE 7-10. The principal pathways for taste.

FROM TONGUE TO BRAIN

Fibers serving the tongue do not form a single nerve. Those from the anterior part of the tongue make up the chorda tympani (*chorda,* string; *tympanos,* drum) branch of the facial nerve. Fibers from the posterior part of the tongue join the glossopharyngeal (*glosso,* tongue; *pharynx,* throat) nerve. Some fibers from the pharynx and larynx become part of the vagus nerve. Fibers from the glossopharyngeal and facial nerves terminate in the solitary nucleus in the medulla where they synapse with neurons that cross the brainstem and ascend via the medial lemniscus to terminate in the bow-shaped arcuate (*arcus,* bow) nucleus. There they synapse with neurons that terminate in the area for body sensation in the parietal region of the cerebral cortex. Figure 7-10 shows the principal tracts involved.

OLFACTION: BY ANY OTHER NAME . . .

The receptors for the sense of smell are located in the upper part of the nasal cavity in the olfactory epithelium, a moist, mucous area where gases can go into solution. These receptors are ciliated bipolar cells (Fig. 7-11) that synapse in the olfactory bulbs at the base of the brain. From here they travel to several cortical and subcortical areas of the brain, particularly to the prepyriform cortex at the base of the brain. The ramifications of the olfactory connections are complex and as yet not

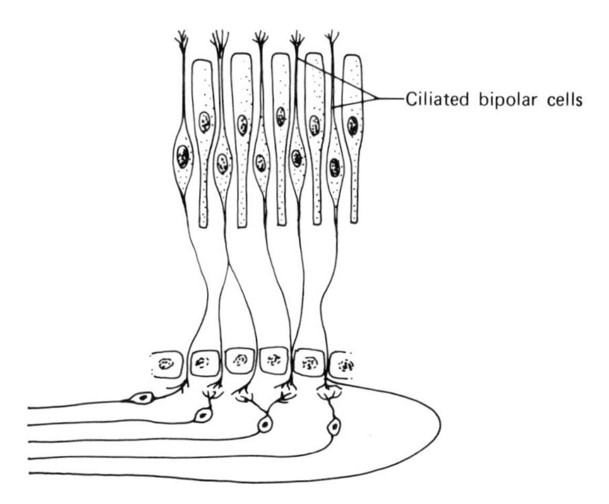

FIGURE 7-11. The ciliated bipolar cells of the olfactory epithelium.

completely understood; therefore, they will not be discussed further here.

The question of what stimuli trigger particular odor sensations has baffled researchers for over three-quarters of a century. Today the most widely accepted explanation is a stereochemical theory, one variety of which has been elaborated by a chemist, John Amoore, who believes that there are seven primary odors, each of which is associated with a molecule of a specific shape. The odors are: camphoraceous, musky, floral, peppermint, ethereal, pungent, and putrid. Camphoraceous, for example, is said to have a bowl-shaped molecule and presumably fits into a bowl-shaped depression in the molecules of the appropriate receptor. Ethereal substances fit an oblong depression. Similarly, each of the primary categories has its particular molecular shape according to the stereochemical theory.

Not all investigators have agreed with the stereochemical theory of smell, since they have found that substances with certain shapes do not smell as they should as predicted by their molecular shape. Moreover, recordings from single fibers leading from smell receptors fail to support such a specific set of receptors for seven primary sensations. There may be as many as fifty primary sensations of smell, since individuals may be odor-blind to this number of separate stimuli. Therefore, the precise nature of the receptor types for olfaction and the primary sensations to which they respond remains a question for future research.

8
THE CRANIAL NERVES

In introducing the nervous system in Chapter 1, we utilized the concept of encephalization of function to characterize the increasing importance of the brain as we go up the phylogenetic scale. It seems appropriate to reemphasize encephalization of function as we begin our study of the cranial nerves that arise directly from the brain. Because the head contains all the sense organs for detecting distant stimuli, a number of these nerves are directly concerned with the special sense organs of exteroception. But at the same time, the skull is covered over with the mass of muscles that we use in facial expression—that most important component of body language—in masticating food, and, if we know the secret, even wiggling the ears. The motor components of some of the cranial nerves control these muscles as well as those of the larynx and pharynx that are used in talking, swallowing, and other oral activities.

If all of this were not enough, the cranial nerves must also mediate cutaneous sensation—touch, pressure, pain, and temperature—for the head region, since the spinal nerves with their large sensory components do not reach to this level of the body. Finally, we shall discover that there is one special cranial nerve that wanders down through the thorax and abdomen as far as the colon giving off branches for the motor control of the thoracic and abdominal viscera.

"ON OLD OLYMPUS'S TOWERING TOP" . . . A QUESTION OF NAMES AND REMEMBERING

The cranial nerves have both numbers and names. Perhaps as a reflection of their importance, they were numbered in the stately Roman system from I to XII, beginning with the first pair that emerges at the most anterior part of the brain and ending with the twelfth pair that emerges at the level of the lower medulla (see Fig. 8-1). Their names, too, reflect the classical scholarship of the anatomists who first discovered them, giving each a Greek or Latin designation in an attempt to

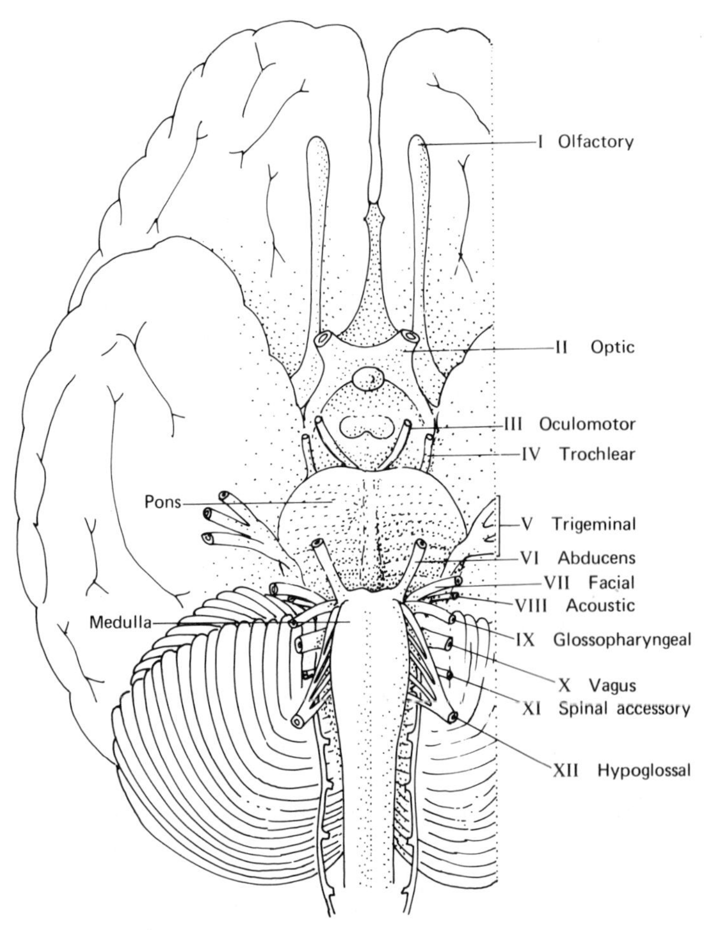

FIGURE 8-1. The under surface of the brain showing the emergence of the cranial nerves.

describe either the appearance or function of the nerves. As a first step in becoming acquainted with these nerves, let us list them by number, name, derivation of name and organ system for which they function. These are presented in Table 8-1.

Upon first acquaintance the names of the cranial nerves seem formidable and difficult to remember, and it must be admitted that this is a traditional hurdle for students of neurology. Somewhere in the dim past a hard-pressed student with a flair for doggerel left as his heritage to scholarship a little verse, which is recommended as a mnemonic aid for remembering the names of the nerves. It goes like this:

> On Old Olympus's Towering Top,
> A Finn and German Vaulted and Hopped.

The secret, if it is not obvious, is that the initial letters of the words in the verse are the same as the initial letters of the cranial nerves. Such are the uses of mnemonic aids: that many former students of neurology can no doubt remember the verse but not the nerves.

SOMETHING OLD, SOMETHING NEW, SOMETHING BORROWED: HOW THE NERVES ARE CLASSIFIED

Anatomists not only name and describe structures but also, in common with other scientists, classify the material with which they work. You will recall the simple classification of motor and sensory nerves and their roots in the spinal cord. There was also the anatomical and functional classification of the autonomic nervous system into the sympathetic and parasympathetic divisions. Both systems were relatively easy to classify, presenting no great challenge to the early anatomists and physiologists. However, the cranial nerves, highly specialized as they are and containing a variety of components, presented far more of a challenge. Because of this complexity, the pioneer anatomists used a combination of functional classifications along with borrowings from embryology and comparative anatomy. And so they established a classification system that, while not always consistent, has become traditional.

We may approach the problem of classification by pointing out that some cranial nerves are purely sensory in function and some are motor, unlike their spinal counterparts, which are always mixed. But some of the cranials are also mixed in function. To complicate matters still

TABLE 8-1
The Cranial Nerves

Number	Name	Derivation of name	Function
I	Olfactory	*olfacere,* to smell	Smell
II	Optic	*optikos,* vision	Vision
III	Oculomotor	*oculus,* eye; *motor,* motion	Movement of the eyeballs, pupillary constriction, and accommodation of the lens
IV	Trochlear	*trochlea,* pulley	Control of eyeballs, the tendon for this muscle functions as a pulley
V	Trigeminal	*trigeminus,* three twins or branches	Sensation from face and head, control of muscles of jaw
VI	Abducens	*abducere,* to draw out	Outward movement of the eyeball

VII	Facial	*facies,* face	Sensory and motor innervation for the face
VIII	Acoustic also statoacoustic or vestibuloacoustic	*akoustikos,* to hear	Hearing and equilibrium
IX	Glossopharyngeal	*glossa,* tongue; *pharynx,* throat	Motor to the salivary glands, sensory to tongue and pharynx
X	Vagus	*vagari,* wandering	Sensory and motor functions of the pharynx, larynx, and thoracic and abdominal viscera
XI	Spinal accessory	*accessori,* a supplement	Sensation from external ear, muscles of pharynx, larynx, and neck
XII	Hypoglossal	*hypo,* below; *glossa,* tongue	Motor control of the tongue

further, some mediate or control general functions, while others are specialized. Finally, some are associated with visceral processes, and others control somatic functions.

In trying to put all of this together, neuroanatomists have established the following basic classes of cranial nerves:

Classification	Functional Example
AFFERENT TYPES	
General somatic afferent	Sensation from skin of face and head
General visceral afferent	General sensation from the tongue and pharynx
Special somatic afferent	Vision and hearing
Special visceral afferent	Taste
EFFERENT TYPES	
General visceral efferent	Muscle regulation and glandular secretion in visceral organs
Special visceral efferent	Control of muscles of mastication and facial expression
General somatic efferent	Control of eye movements

At first reading, the traditional system of classification may appear confusing rather than helpful as such a system ought to be. The classification of the afferent components seems reasonable enough, but why, for example, would the muscles of mastication and facial expression, which are striated and under voluntary control, be classed as "visceral," while muscles of the eyeballs, which are also striated and under voluntary control, be classed as "somatic"? The answer lies first in the tradition or custom of comparative anatomists of classifying those muscles and sensory systems having to do with the alimentary canal or respiratory system as visceral. Second, those nerve components that innervate muscles derived from the embryonic branchial arches are also classified as visceral. Branchial arches occur in the neck region of the embryo. In amphibian forms the gills and mouth develop from these arches. The human embryo does not develop gills but does develop branchial arches. The muscles that arise from mesodermic tissues in these arches are classed as visceral because of this derivation. Applying this same logic to the cranial nerves, those that control the muscles of facial expression, which are derived from the second branchial arches, are classed as visceral as are the nerves controlling mastication, which are derived from the first arches.

We are ready to grant that a less complicated and more logical system appears to be needed, but traditions are as difficult to modify in neurology as they are in the area of social customs such as weddings. We shall thus try to justify and explain the traditional classification of each of the nerve components as we describe them in detail.

I. Olfactory. The olfactory nerve is classified as a special visceral afferent. It is specialized in that it carries only sensory impulses that arise from the olfactory epithelium. It earns its visceral designation because smell is closely associated with the activities of the gastrointestinal tract and respiration. The origin of the nerve in the olfactory epithelium and its termination in the olfactory bulbs were discussed in Chapter 7.

II. Optic. The optic nerve, as we pointed out in Chapter 7, is more accurately described as a fiber tract of the brain. However, custom has included it in the cranial nerves as a special somatic afferent. We have already traced its origin and termination in Chapter 7 following our discussion of the mechanisms of vision.

III. Oculomotor. The oculomotor nerve is classified as: (1) a general somatic efferent. This component of the oculomotor innervates and controls the extrinsic muscles that move the eyeball (superior, inferior and intermedius recti, and inferior oblique) except for the superior oblique and lateral rectus (Fig. 8-2). The classification "general somatic"

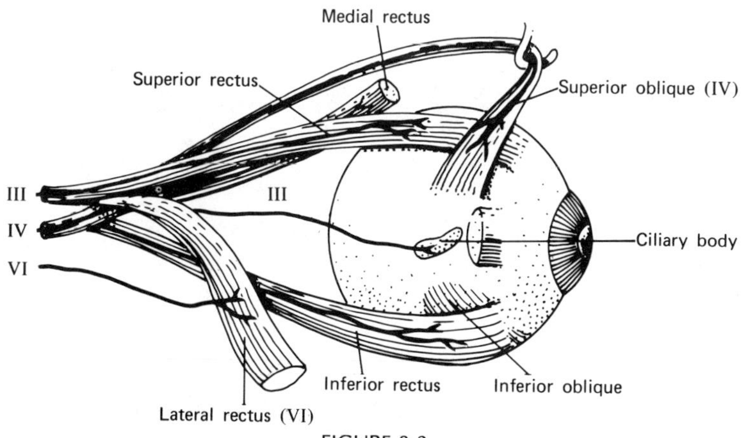

FIGURE 8-2.

refers to the fact that the muscles of the eyeball arise from the dorsal mesoderm in the embryo in contrast to visceral muscles, which arise from the ventral mesoderm. We also know that the muscles of the eyeballs are of the somatic type in that we can control them voluntarily, although much of their action in everyday life is automatic. That this cranial nerve as well as the trochlear and abducens are collectively concerned with movement of the eyes testifies to the importance of these high-speed and marvelously coordinated muscles. (2) The second component of the oculomotor is a general visceral efferent component consisting of parasympathetic fibers that innervate the sphincter of the iris and the ciliary muscle for light adjustment and accommodation. (3) The third component is a general somatic afferent component carrying kinesthetic impulses from the eye muscles under (1).

IV. Trochlear. (1) The trochlear is classed as a general somatic efferent nerve controlling the superior oblique muscles of the eyeball for outward and downward movement. (2) The second component is a general somatic afferent carrying kinesthetic impulses from the superior oblique muscles.

V. Trigeminal. Arising under the base of the brain, the trigeminal forms three branches, the ophthalmic, maxillary, and mandibular (Fig. 8-3). It contains nerve fibers that are: (1) a general somatic afferent

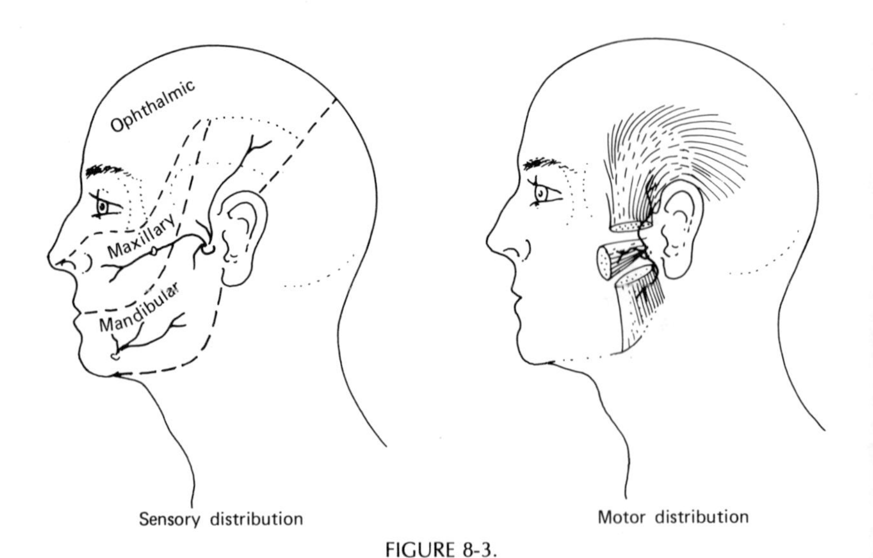

FIGURE 8-3.

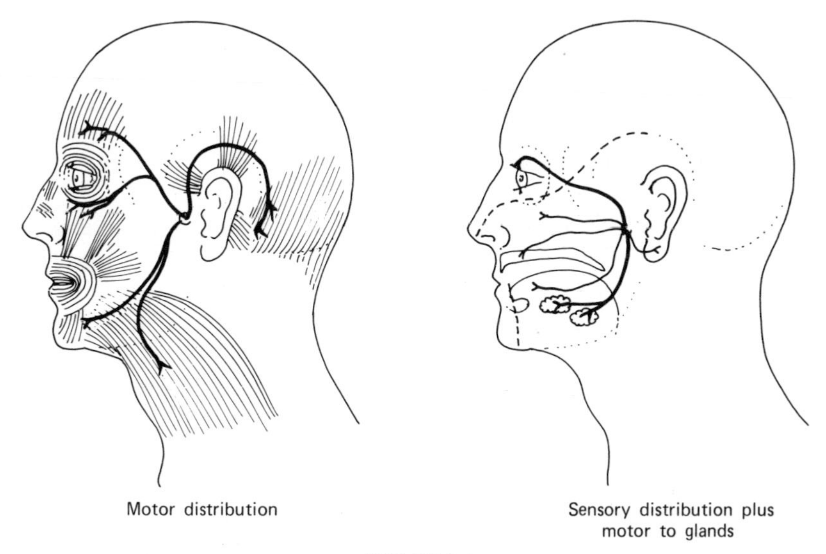

FIGURE 8-4.

mediating cutaneous sensation from the face and head; (2) a special visceral efferent from the masticator nerve to the muscles of mastication controlling the lower jaw; (3) a general somatic afferent consisting of kinesthetic fibers from the jaw muscles. Note, again, the designation "visceral" is used even though the jaw muscles are voluntary and striated. This borrowed classification recognized the close associaton of this component of the trigeminal with the evolutionary and embryonic development of the gastrointestinal tract.

VI. Abducens. (1) A general somatic efferent to the lateral recti muscles of the eyeball for outward rotation. (2) A general somatic afferent component from the lateral recti muscles for kinesthesis (see Fig. 8-2).

VII. Facial. This complex nerve has both sensory and motor components. The sensory components are: (1) a general somatic afferent from part of the external ear mediating cutaneous sensation; (2) a general visceral afferent mediating deep sensitivity from the mucous membranes of the nose and proprioceptive impulses from the muscles of the face; (3) a special visceral afferent mediating taste from the anterior two-thirds of the tongue.

The motor components of the facial nerve are: (1) general visceral efferent fibers of the parasympathetic innervating the lacrimal glands of the eye and the submaxillary and sublingual salivary glands; (2) special

CASE VIII

TRIGEMINAL NEURALGIA:
THE INTOLERABLE PAIN

Miss T.N., a twenty-six-year-old nurse, was admitted to the medical center following the sudden onset of severe facial pain. She described the pain as spasmodic, so severe it was "like an acetylene torch" in front of her face and behind her left eye. The accompanying sketch shows the areas of her face affected. The bouts of pain lasted from 5–15 minutes, during which time she was immoblized by the agony.

Three weeks prior to the onset of her symptoms, T.N. had undergone extensive dental work for abcesses that were treated by antibiotics and extraction. Ten days later the pain in her face began. At first it was triggered by bright light, but in succeeding days a breath of wind, a light touch on the face, or contact inside the cheek brought on spasms. In an attempt to lessen the frequency of the attacks, T.N. began to wear an eye patch over her left eye.

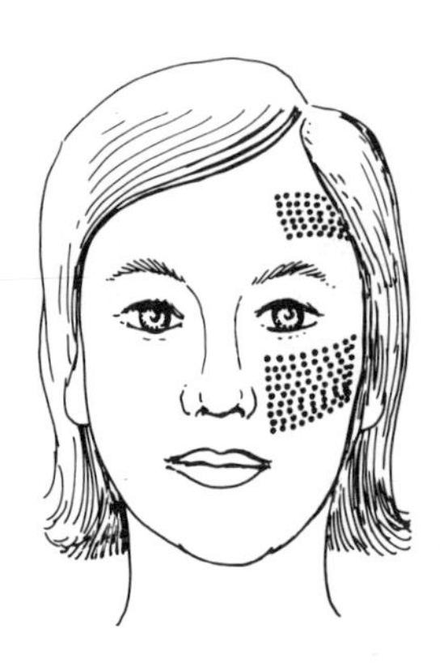

Neurological examination led to the diagnosis of trigeminal neuralgia, of probable traumatic origin. Extensive treatment with every available drug was tried until the pain could no longer be controlled, even with large doses of narcotics. So severe were her symptoms, she was unable to eat and had to be fed intravenously.

This somewhat unorthodox regimen was prolonged rather than electing immediate surgical intervention, since T.N.'s familiarity with the disorder led her to fear motor paralyses; loss of protective tearing with possible ulceration of the cornea; anesthesia in the face, gums, and teeth; and loss of control of the jaw muscles with trauma from self-inflicted bites while eating.

The patient was prevailed upon to undergo a cocaine block in order to demonstrate the effectiveness of relief for a prolonged period of interrupting the middle branch of the trigeminal and therefore convince her of the desirability of surgery.

The success of the block convinced T.N. of the desirability of immediate surgery, and under general anesthesia the middle division of the trigeminal was exposed and cut with both ends being rolled back and removed for about 2.5 centimeters. At the cranial end a cautery was introduced into the stump and the nerve coagulated to prevent possible regeneration. Following recovery from surgery the patient was free of pain and discharged without any serious side effects from the operation.

visceral efferent fibers innervating the superficial muscles of the face, scalp, and ear—the muscles for facial expression. This component of the facial is classified as visceral since the muscles that they serve arise from the second branchial arch in the embryo, which is associated with the development of the gills in lower forms. Because gills in fishes and amphibians function in respiration, they are classified as visceral. This designation has been borrowed and taken over into the classification of components of the facial nerve that serve the muscles whose origin is the mesoderm of the second branchial arch. From a *functional* point of view, the muscles act as somatics.

VIII. Acoustic. The acoustic, or vestibuloacoustic, is a nerve with two main divisions, the cochlear and vestibular, reflecting the separate auditory and labyrinthine functions of the nerve.

The cochlear is classified as a special somatic afferent component. This component mediates hearing. The vestibular component is also classified as a special somatic afferent mediating labyrinthine sensitivity. Both of these special senses are discussed in detail in Chapter 7.

IX. Glossopharyngeal. Like the facial, this nerve is a complex one serving a number of different functions. The sensory components are classified as: (1) general somatic afferents mediating cutaneous sensation from the region of the ear; (2) general visceral afferents mediating general sensation from the tongue and pharynx; (3) special visceral afferents mediating taste from the posterior third of the tongue. (4) There are also special visceral efferents controlling the pharyngeal muscles (see Chapter 7).

X. Vagus. The vagus is not only the longest of the cranial nerves, traveling as it does as far down as the colon, it is also a complex nerve mediating five separate types of functions: (1) a general somatic afferent component mediating cutaneous sensation from the external ear; (2) the general visceral afferent component mediating general sensation from the pharynx, larynx, and thoracic and abdominal viscera; (3) special visceral afferent fibers mediating taste from the epiglottal taste buds; (4) general visceral efferent components, which are the parasympathetic preganglionic fibers to the thoracic and abdominal viscera, regulating the secretion of visceral glands, cardiac muscle, and the muscles of the gastrointestinal tract; and (5) special visceral efferent components innervating skeletal muscles of the palate, larynx, and pharynx—visceral because of their association with the gastrointestinal tract (see Figure 5-1).

XI. Spinal accessory. A helper for the vagus and cervical spinal nerves, this nerve (Fig. 8-5) contains: (1) general somatic afferent components that are proprioceptive fibers from the muscles of the neck,

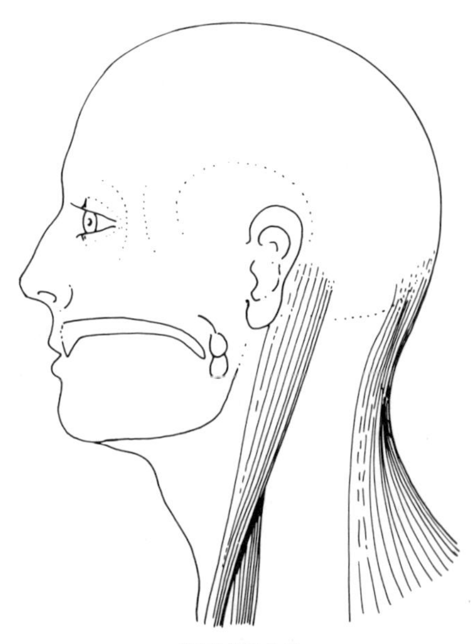

FIGURE 8-5.

which are controlled by the motor fibers of this nerve; (2) general visceral afferent fibers that, with the vagal cardiac branches, regulate heart beat and other thoracic functions; and (3) special visceral efferent fibers to the pharynx and larynx, with a branch that serves two neck muscles, the trapezius, and the sternomastoid. It should be noted that this branch has its cell bodies in the anterior gray matter of the first five segments of the cervical cord, indicating that this component of the nerve is truly a part of the spinal system both anatomically and functionally.

XII. Hypoglossal. A general somatic efferent nerve, the hypoglossal (Fig. 8-6) controls the strap muscles of the neck and tongue. It is likely that there are also general somatic afferent components in this nerve from the same muscles mediating kinesthesis, but their points of origin and termination are uncertain.

AN AFTERTHOUGHT AND A LOOK FORWARD

The complexity of the cranial nerves with their diverse anatomical ramifications and functional components has reached its highest point in the primates. However, it is interesting to note that homologues (*homologos,*

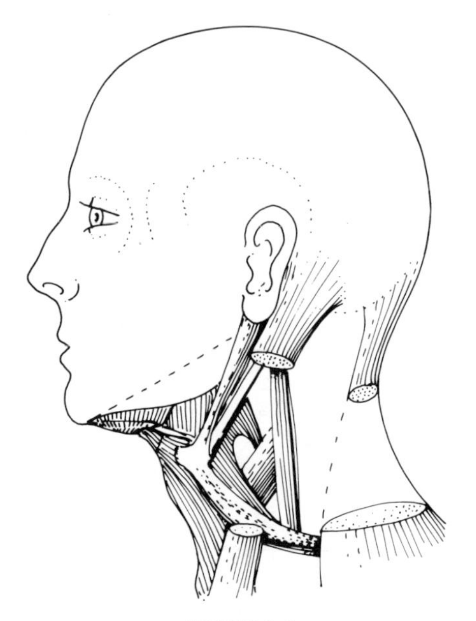

FIGURE 8-6.

agreeing—here in structure and origin) for the cranial nerves of the primates can be found in *amphioxus,* a lancelet and so-called connecting link between vertebrates and invertebrates. In the fishes, which possess a true vertebral column (unlike amphioxus, which has only a notochord), ten of the cranial nerves characteristic of the primates can be identified. Between fishes and man few changes in individual cranial nerves occur, except the addition of the eleventh and twelfth in the amniotes or animals whose embryos are enclosed in an amniotic sac. Thus, from a comparative point of view, the cranial nerves are among the most constant of the vertebrate organs.

In the following two chapters we shall be meeting them again as we discuss their sites of origin or termination in the medulla and pons.

9
THE MEDULLA OBLONGATA

A MIGHTY INCH

Breathing, gagging, coughing, swallowing, vomiting, and articulating sound and song—all these are the evolutionarily ancient functions entrusted to the inch of brain just above the spinal cord that lies inside the skull. Called the medulla oblongata (*medulla*, marrow; *oblongata,* oblong), it scarcely looks different from the spinal cord to which it is connected. The vital functions of the medulla were known to Galen (A.D. 130–200), the Greek physician who sliced through it paralyzing the respiratory centers in an animal, and, incidentally, became one of the first investigators to study the functions of the brain by removing parts and noting the effects. The modern physician also knows how vital the medulla is. Carefully drawing a sample of spinal fluid from the vertebral column for diagnostic purposes, he or she occasionally creates too much negative pressure and the brain drops downward squeezing the medulla against the foramen magnum or large opening into the floor of the skull. The patient suddenly stops breathing. Tumors, too, can also produce a "pressure conus," as the condition is called, with the sudden death of the victim from respiratory failure.

The medulla is more than a vital center, important as its respiratory and other regulatory functions are. It is also a trunk line for the fibers of the afferent and efferent systems that travel up and down from the brain. All the pathways that we identified in Chapter 3 must go through the medulla. Here, too, a number of cranial nerves have their nuclei of origin or termination. These are scattered in among the ascending and descending clumps of gray matter. Finally, this complex little structure is part of the reticular formation, a network of intermixed white and gray matter that extends up through the brainstem to the cerebral cortex. Its function, broadly speaking, is that of alerting the cortex and determining which of the millions of stimuli streaming up from the cord will be allowed to go on to the cortex and which will not. We shall be dealing with the reticular formation in detail in Chapter 11. In the present chapter we shall concentrate on the conducting and regulatory functions of the medulla.

A THING OF SHREDS AND PATCHES

If the brain is freed from the skull by cutting the cord and cranial nerves, the external appearance of the medulla is not prepossessing. It is similar in size to the spinal cord, except that it swells outward near its junction with the pons to assume a roughly pyramidal shape. Hanging from it are the shreds of the cranial nerves. If we look more closely at the medulla from a dorsal or posterior perspective (Fig. 9-1), we see the familiar posterior median sulcus that is continuous with the dorsal sulcus in the

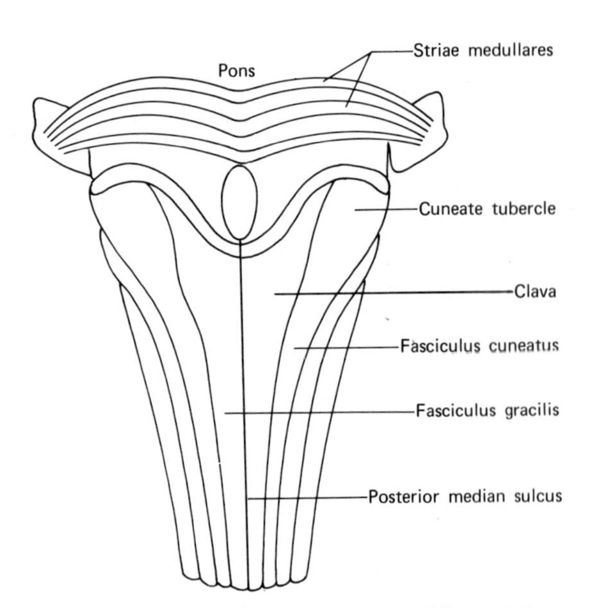

FIGURE 9-1. Dorsal or posterior view of the medulla.

spinal cord. On either side of the midline, there are low ridges as if patched on as an afterthought. These are the large tracts of fasciculi gracilis and cuneatus on their way upward to end in their nuclei of termination in the upper medulla. The site of the nuclei can be distinguished externally as small swellings that are known as the clava (*clava,* club) overlying the nucleus of gracilis, and the cuneate tubercle (*cuneus,* wedge-shaped; *tuberculum,* a nodule or swelling) over the underlying nucleus of cuneatus. The upper limit of the medulla is marked off by the striae medullares (*stria,* furrow, channel), transverse stripes that belong to the posterior limit of the pons.

From the ventral or anterior aspect (Figure 9-2), the medulla is divided along the midline by a longitudinal groove called the anterior median fissure. On either side of the fissure are ridges known as the pyramids, formed by the pyramidal tracts containing the corticospinal fibers that descend to the lower motor neurons. The crossing or decussation of the lateral corticospinal tracts may be seen near the caudal end of the medulla just before the region where it joins the cord.

Arising out of the ventral medulla are the fibers of the cranial nerves.

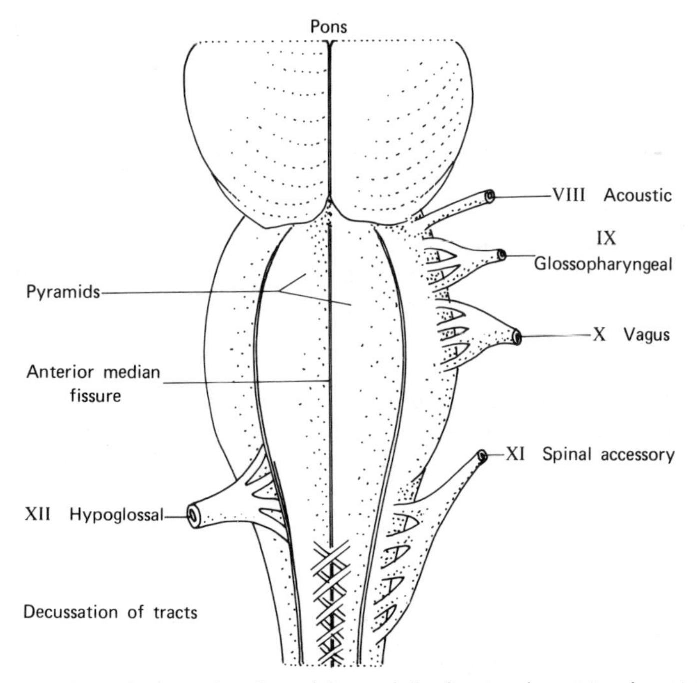

FIGURE 9-2. Ventral or anterior view of the medulla showing the origin of cranial nerves VIII–XII.

(VIII, IX, X, XI, and XII). From the upper level of the medulla the eighth, or acoustic, nerve fibers come in from the organ of Corti, the semicircular canals, vestibule, and saccule.

The ninth, or glossopharyngeal, nerve emerges just below the acoustic, its mixture of sensory components mediating taste from the posterior tongue and general sensation from the ear are mixed in with outgoing motor fibers on their way to the parotid gland and muscles of the pharynx.

The tenth, or vagus, arises as a number of rootlets emerging from the lateral medulla. Some of these form the glossopharyngeal nerve but most form the vagus, which supplies motor and sensory fibers for the mucous membranes of the pharynx and larnyx. In addition, the vagus is widely distributed to the organs of the thorax and abdomen mediating parasympathetic functions.

The eleventh, or spinal accessory, has two distinct roots, as we pointed out in Chapter 8. The spinal root has its origins in the anterior horn cells of the first five cervical segments of the spinal cord. These fibers join together and ascend through the foramen magnum to blend with the cranial portion of the nerve. After accompanying the spinal root for a short distance, the fibers from the cranial branch diverge again to join the vagus for distribution to the thoracic and abdominal viscera. The spinal fibers, it will be recalled, innervate the sternomastoid and the trapezius muscles of the neck.

The twelfth, or hypoglossal, is the most caudal of the cranial nerves, arising anterior to the decussation of the pyramids. Its peripheral branches leave the skull through an occipital opening to innervate the strap muscles of the neck and tongue.

From the lateral aspect (Fig. 9-3), the medulla appears as a bulblike structure continuous with the spinal cord caudally and blending into the mass of the pons rostrally. There are two sulci, the dorsolateral and ventrolateral. The rows of rootlets of the cranial nerves arise along the more rostral limits of these sulci. The prominent swelling between the sulci is the olive, marking the site of an underlying cluster of nuclei known as the olivary complex, whose functions will be taken up in the following section.

A SCIENTIFIC GUILLOTINE: MICROSCOPIC SECTIONS THROUGH THE MEDULLA

Even an inch of the brain, such as the medulla, is far too complex to understand unless many microscopic cross sections at different levels are

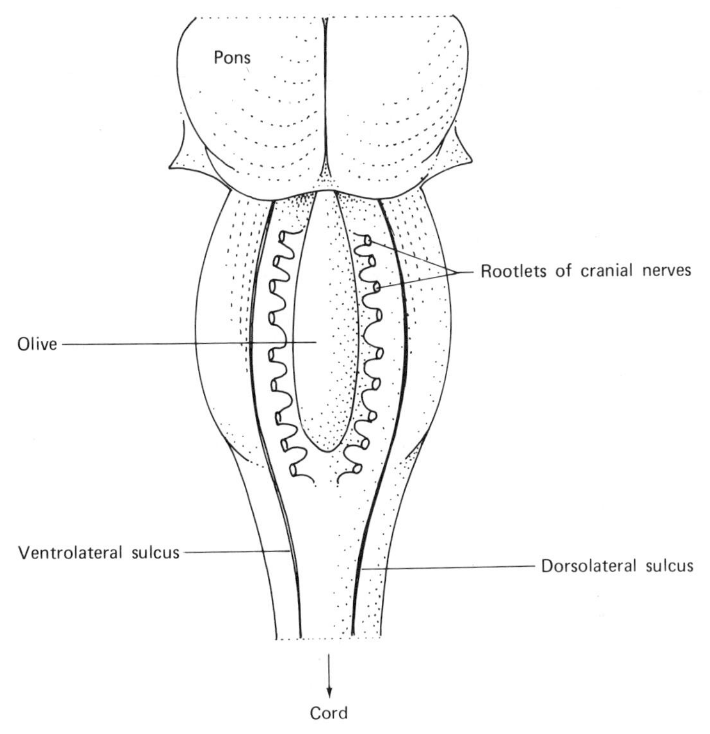

FIGURE 9-3. Lateral view of the medulla.

prepared in order to search out and visualize the location and distribution of nuclei and fiber tracts. The neuroanatomist's guillotine is the microtome, a knife used to understand life, not to destroy it. The microtome can slice suitably prepared tissues into sections a few microns thick, which can be stained by various methods either to bring out cell bodies or fiber tracts clearly. When properly mounted on slides, a series of such sections helps the anatomist to reconstruct the appearance of the tissues from the inside, so to speak. Some of the most honored men in the history of neuroanatomy, Camillo Golgi (1843–1926), Ramon y Cajal (1852–1934), and Karl Weigert (1845–1904), dedicated their professional lives to the development of staining techniques that make it possible to study the anatomical properties of neurons and their interrelationships.

Sections through the medulla taken at progressively ascending levels reveal changes in both the appearance and spatial relationships of fiber tracts as they course through the nuclei that are interspersed at all levels of the medulla. Figure 9-4 shows diagrammatically the appearance of the medulla at the level of the decussation of the pyramidal tract. Its appear-

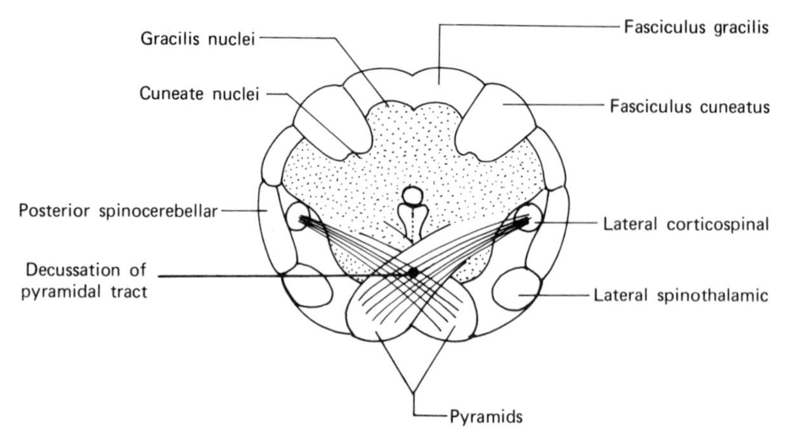

FIGURE 9-4. A cross section of the ower medulla showing the decussation of the pyramids.

ance at this level is not dissimilar to that of the spinal cord. The tracts of gracilis and of cuneatus mediating conscious kinesthesis and pressure sensitivity are prominent in the dorsal region, and the butterfly-shaped gray matter of the central cord is also clearly evident in the medulla. Several of the tracts with which we became familiar in Chapter 3 can be readily identified and are shown in Figure 9-4.

A cross section taken about one-half way between the beginning of the medulla at the spinal level and its termination at the pons reveals a number of changes (Fig. 9-5). The tracts no longer retain the same spatial relationship with each other as they had at lower levels. The ventral corticospinal tracts are still quite prominent, and some of the familiar ascending tracts, such as the unconscious kinesthetic ventral spinocerebellar and the lateral spinothalamic mediating touch, can be identified, but the dorsal tracts of gracilis and of cuneatus, which ended at the nuclei of gracilis and of cuneatus, respectively are no longer present. The most prominent features of the middle medulla are the following. First, the bilateral olivary nuclei are easily distinguished by their pleated shape. These nuclei receive collateral fibers from the descending pyramidal tracts. Second, they synapse here with fibers that go on to the cerebellum. These connections are one of the brain's sites for the coordination and integration of activities of the motor executive areas of the cerebral cortex with the coordinating centers of the cerebellum.

Third, the medial lemniscus, a band of fibers mediating proprioceptive impulses arising from the nuclei of gracilis and of cuneatus to their point of termination in the ventral posterolateral (VPL) nucleus of the thalamus, becomes a prominent feature of the middle medulla.

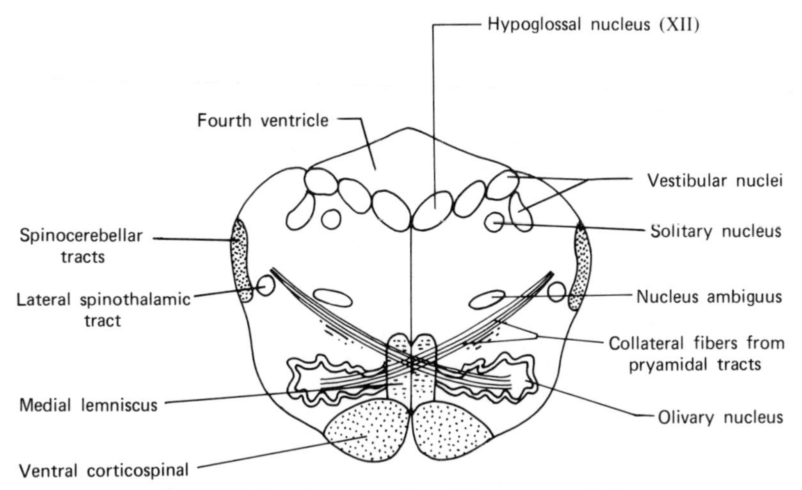

FIGURE 9-5. A cross section of the upper medulla.

Fourth, several nuclei of cranial nerves may be distinguished in this region. The nucleus ambiguus, so named because of its diffuse structure, is a collective motor nucleus for the somatic components of the glossopharyngeal, vagus, and spinal accessory nerves. It lies ventromedial to the nucleus of the hypoglossal nerve. The nucleus of the solitary tract lies just dorsal to the nucleus ambiguus. It is the nucleus of termination for incoming afferent root fibers of the vagus nerve. The vestibular nuclei can also be readily distinguished in the middorsal region of the medulla at this level. It is one of the nuclei of termination of the vestibular branch of the eighth nerve and forms part of a complex of vestibular nuclei that receive fibers from the vestibular system.

Finally, the reticular formation constitutes a large portion of the gray and white matter of the medulla. It is part of a complex of cells extending up into the pons and brain proper. The functions of these cells will be discussed later.

THE LONG VIEW

It is also possible to visualize the cranial nerve nuclei of the medulla from a longitudinal point of view. These are shown in Figure 9-6 with their nerve numbers. It should be noted that some nuclei are not confined to a small portion of the medulla but extend some distance up and down, an arrangement that is demanded when neural centers must receive or give off fibers at different levels.

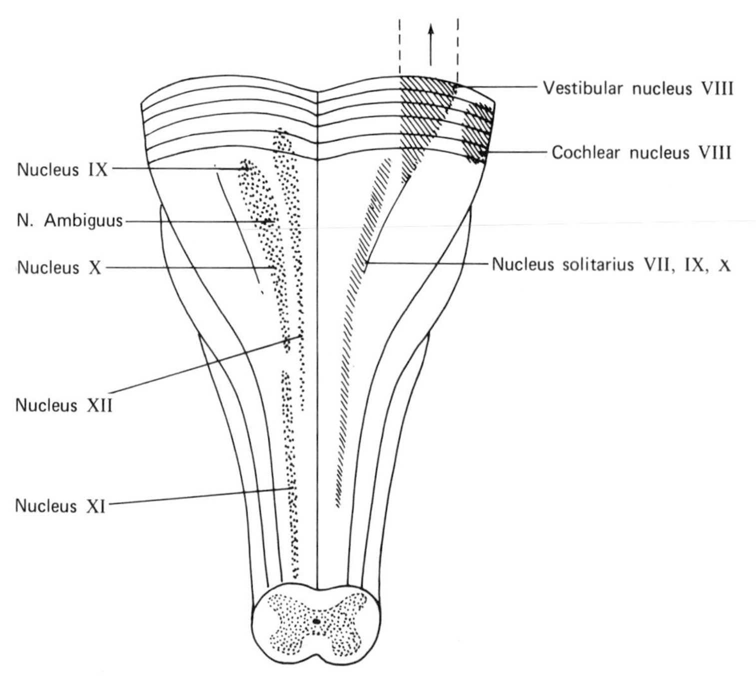

FIGURE 9-6. The nuclei of the medulla shown from a longitudinal perspective in a supposedly transparent view from the dorsal surface. For clarity the motor nuclei are shown on the left side (stippled) and the sensory nuclei on the right side (cross-hatched).

THE BREATH OF LIFE: RESPIRATORY REGULATION IN THE MEDULLA

The respiratory center—more accurately a collection of centers—is a group of neurons dispersed through the reticular formation in the medulla and pons (Fig. 9-7). The respiratory center of the medulla is also known as the medullary rhythmicity center, since neurons in this region control the rhythmic cycle of inspiration-expiration. This can be demonstrated in experimental animals by transecting or cutting the cord immediately above and below this center. The basic rhythmicity of inspiration and expiration continues. However, it is much weaker than it is in the intact animal showing that contributions from other areas are also involved in its normal functioning.

One of these is the contribution from afferent neurons from the spinal cord. If the medulla is transected just above the respiratory center

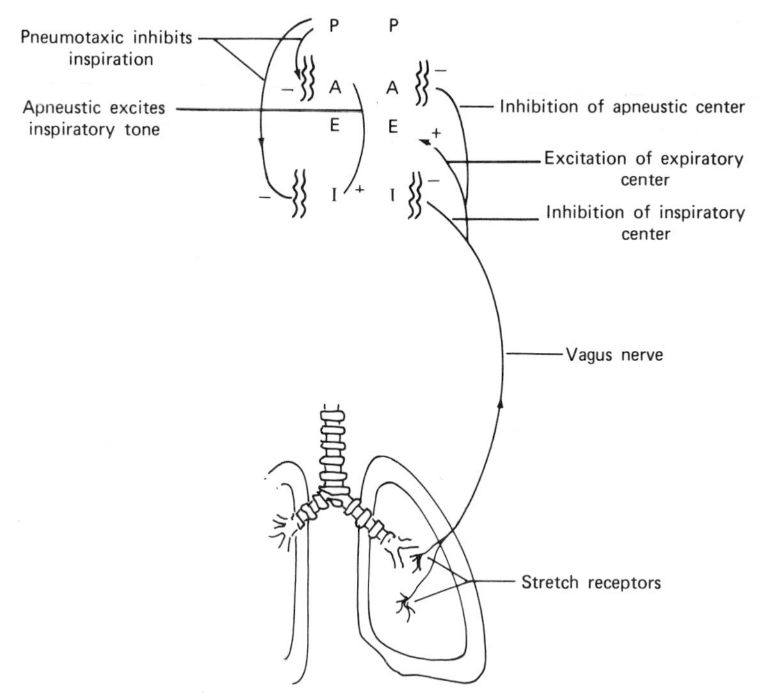

FIGURE 9-7. The respiratory centers of the medulla and pons. The apneustic center normally excites the inspiratory center of the medulla. The pneumotaxic center inhibits the apneustic center or the inspiratory center of the medulla directly.

but the cord is left intact, respiration occurs in gasps, indicating an imbalance of afferent impulses arriving at the center from the thoracic region. These impulses are generated by stretch detectors in the lungs that help regulate the extent of inspiration and expiration (the Hering-Breuer reflexes). Excessive facilitatory signals from the spinal cord are normally balanced by impulses from the vagus and glossopharyngeal nerves and from the cerebral cortex and midbrain. Clearly, the basic inspiration-expiration rhythm generated by pools of cell bodies in the medulla is but one part of an extremely complex system.

The complexity of the respiratory cycle is revealed by the discovery that there are two additional centers located above the medulla in the pons for regulation of respiration. They are the apneustic and pneumotaxic centers. The function of the apneustic center is observed when the pons is transected between the apneustic and pneumotaxic areas. The animal's breathing pattern is one of prolonged inspiration but shortened expiration. Leaving the pneumotaxic center intact allows for a normal balance between inspiration and expiration. Thus it would appear that the pneumotaxic center functions to maintain control over the inspiratory area in the apneustic region of the pons and also helps to regulate a balanced rhythmicity of the medullary center. So much depends upon so little.

10 THE PONS AND THE MIDBRAIN

A BRIDGE OF MANY CROSSINGS

Located between the medulla and cerebellum, a massive bundle of nerve fibers makes its crossing connecting the two cerebellar hemispheres—the right half must know what the left is doing. Named, the pons (*pontis*, bridge), appropriately enough, this structure also contains fiber tracts entering or leaving the brain, nuclei for the cranial nerves, and components of the reticular activating system. A short step rostrally brings us to the midbrain or mesencephalon, the less than an inch of the brain that connects the pons to the cerebrum, and like the pons, with which it is continuous, the midbrain is a rich complex of tracts, nuclei, and fibers belonging to the reticular system. Our task in this chapter will be to summarize the more important anatomical and functional components of these two divisions of the brain.

THREE VIEWS OF THE BRIDGE

Looking up under the base of the brain as one might view the underside of a bridge while passing under it on the river below, the anterior or ventral pons appears as a large bulbous swelling (Fig. 10-1). A shallow furrow or sulcus marks the midline division. A blood vessel, the large

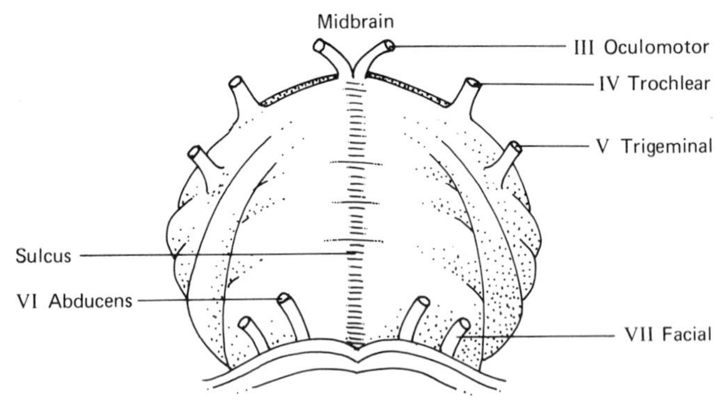

FIGURE 10-1. The pons from the ventral or anterior aspect, showing the origin of cranial nerves III–VII.

basilar artery, runs along this furrow nourishing the pons, cerebellum, and midbrain. At the junction of the pons and midbrain, the oculomotor and trochlear nerves emerge on their way to the eye muscles. The large trigeminal nerve serving the face appears at the midpons, and the abducens, for sideways glances of the eyes, at the junction of the pons and medulla.

Turning the brain on its side with the cerebellum removed gives a good lateral view of the pons (Fig. 10-2). Again it appears as a bulblike

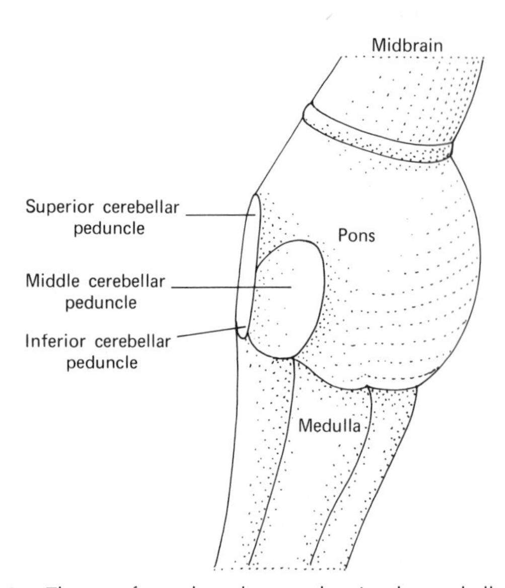

FIGURE 10-2. The pons from a lateral aspect showing the cerebellar peduncles.

swelling between the medulla and the midbrain. From this aspect the most conspicuous structures are the middle cerebellar peduncles (*peduncle*, little foot) which connect the pons to the cerebellum. The cerebellar peduncle is also known as the brachium pontis (*brachion,* arm). Apparently the early anatomists had some difficulty in choosing between an arm and a leg in naming this conducting bundle, but in either case the meaning they were attempting to convey is clear—it is a limb linking the cerebellum and pons. We might also note that the cerebellum is connected to the midbrain by the superior cerebellar peduncle and to the medulla by the inferior cerebellar peduncle. As Figure 10-2 shows, all three peduncles are closely associated anatomically.

From the posterior or dorsal aspect (Fig. 10-3) the top of the pons forms the floor of the fourth ventricle (*ventriculus*, little belly), a cavity filled with cerebrospinal fluid. The body of the pons is seen as a triangular-shaped structure with faint transverse markings, the striae medullares, a band of fibers belonging to the cerebellar system..

SLICING AGAIN: THE INTERIOR OF THE PONS

The relative simplicity of the external pons contrasts sharply with the complexities of the interior structure. And, as we found to be true in the case so the medulla, the appearance of the pons varies with the level at

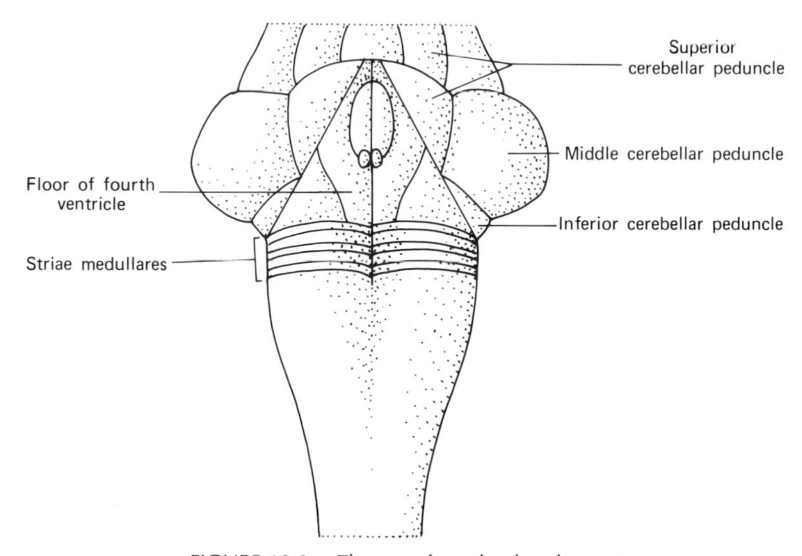

FIGURE 10-3. The pons from the dorsal aspect.

which the section is taken. We shall look at two representative sections, one from the lower and one from the middle region.

At first glance there is a temptation to turn away in bewilderment at the complexity of the interior structures of the lower pons (Fig. 10-4), and something of this feeling must have been the lot of the first anatomists who looked through their microscopes at sections of the brain. Yet a few moments' study reveals that basically the pons is made up of two types of fiber tracts interspersed among scattered nuclei. Looking first at the ventral region, we see numerous transverse fibers that eventually funnel into the tightly packed middle cerebellar peduncle. These neurons consist primarily of axons from the pontine nuclei, which project laterally and which will be discussed later.

Nestled among the transverse fibers are longitudinal tracts prominent among which are the familiar corticospinal tracts on their way to the muscles of the body below the neck and new acquaintances, the corticipontine tracts, which link the cerebral cortex to the pons and then go on to the cerebellum. Closely associated with these tracts are the pontine nuclei, small, dispersed clumps of cell bodies scattered throughout this region. The pontine nuclei function as synaptic centers for the corticopontine fibers.

Just above the region of the corticospinal and corticopontine tracts are the medial lemniscus and trapezoid body. It will be recalled that the ascending proprioceptive tracts of fasciculus gracilis and fasciculus cuneatus form the medial lemniscus after they leave the nuclei of gracilis and of cuneatus in the medulla and cross over to the opposite sides. The trapezoid body, named for its shape, is a collection of auditory relay neurons which forms, in this region, a prominent band of decussating or crossing fibers intermingled with nuclei. The neurons of the trapezoid body originate mainly in the ventral cochlear nucleus. From there, many cross over the trapezoid body to the superior olivary nucleus on the opposite side of the pons, where they synapse with fibers traveling to the auditory area of the cortex.

As we move forward toward the fourth ventricle, we can identify the lateral spinothalamic tract on either side of the medial lemniscus mediating pain and temperature. And above the medial lemniscus the medial longitudinal fasciculi appear as prominent bundles of fibers. Some fibers of the medial longitudinal fasciculi arise from the vestibular nuclei, while others are part of the reticular activating system.

Above the fourth ventricle is a prominent irregular nucleus called the dentate from its resemblance to a set of teeth (Fig. 10-4). It is associated with cerebellar functions and will be discussed further in Chapter 12. Located just below the termination of the ventral arms of the dentate nucleus are the superior and lateral vestibular nuclei, which receive fibers from the vestibular system.

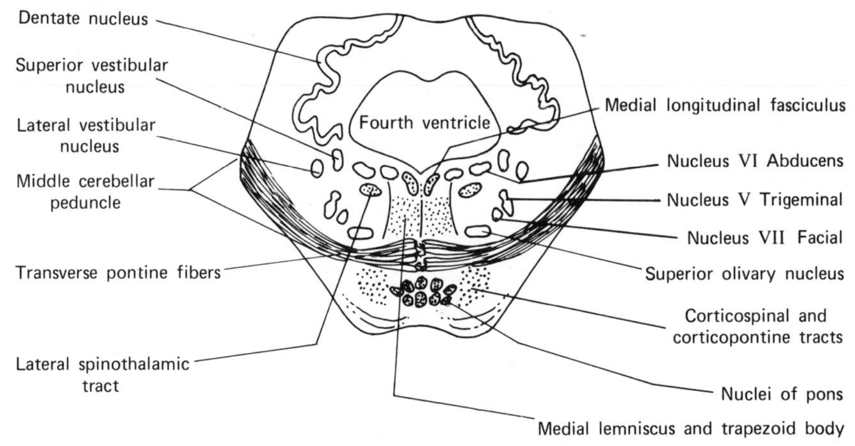

FIGURE 10-4. A cross section of the lower pons showing the interior structures.

Finally, we may identify the pontine nuclei of the cranial nerves. As Figure 10-4 shows, the nuclei of the abducens or sixth nerve lie in the medial portion of the pons just below the fourth ventricle. The nuclei of the facial or seventh nerve lie dorsal to the superior olivary nuclei. The motor nuclei of the trigeminal or fifth nerve and its main sensory nuclei are located close together in the dorsolateral portion of the fibers of the reticular formation.

AT THE MIDDLE OF THE BRIDGE

Figure 10-5 shows the principal tracts and nuclei identifiable in a cross section of the middle region of the pons. Even a superficial examination reveals that no structures are involved with which we are not already

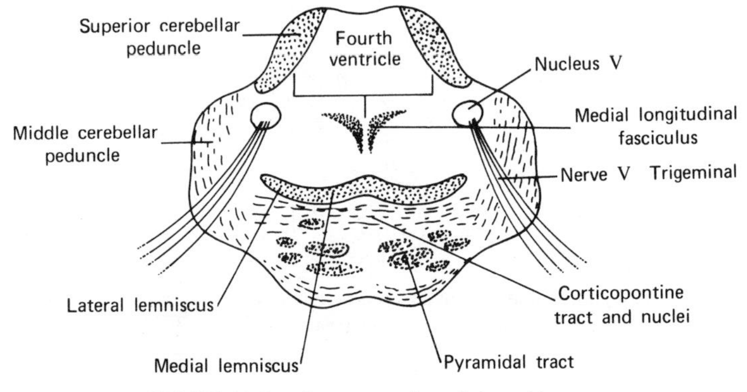

FIGURE 10-5. A cross section of the mid pons.

familiar from our study of the lower level, although the spatial relationships are, to be sure, different. We might also emphasize that some of the same cranial nuclei that appear in a cross section of the lower pons are still evident in the middle region. This is because many of these nuclei extend some distance longitudinally in the midbrain, pons, and medulla (see Fig. 10-6.)

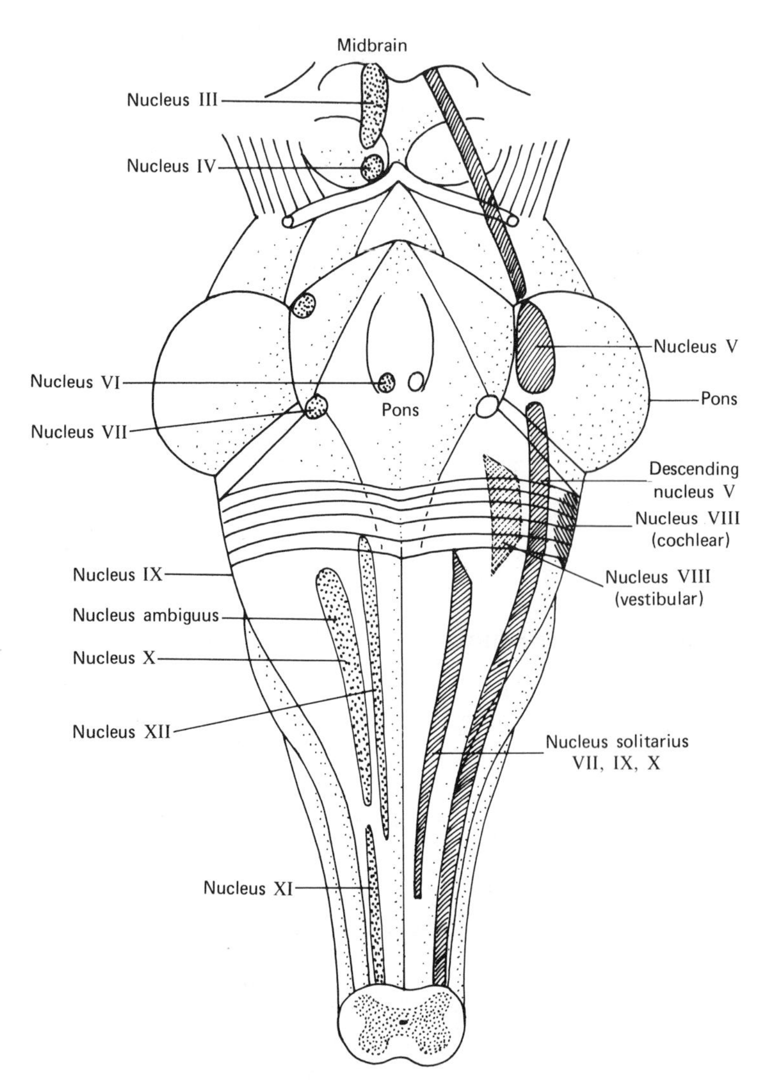

FIGURE 10-6. A longitudinal section of the pons and medulla in a supposedly transparent view from the dorsal aspect. For clarity the motor nuclei are shown on the left (stippled); sensory nuclei are shown on the right (cross-hatched).

THE MIDBRAIN: BUNDLES, BUMPS, RED, AND BLACK

The midbrain is a short segment of the cerebrum that lies between the pons and the structures of the diencephalon. Running through its center is the cerebral aqueduct, a tubular connection between the third and fourth ventricles for the passage of cerebrospinal fluid. The most conspicuous structures on the ventral or anterior aspect of the midbrain (Fig. 10-7) are the large cerebral peduncles, two ropelike bundles of fibers connecting the midbrain to the pons. The dorsal portion of the exterior midbrain is marked by four prominent swellings or elevations known as the corpora quadrigemina (*corpus*, body; *quadrigemina*, four twins). The rostral pair are the superior colliculi; the caudal pair, the inferior colliculi.

A cross section (Fig. 10-8) of the lower part of the midbrain reveals that the inferior colliculi are two large nuclei at the roof (tectum) of the midbrain. The inferior colliculi are concerned with auditory reflexes. They receive fibers from the auditory nerve and project outgoing fibers to the medial geniculate body carrying impulses to auditory centers in the cerebral cortex. It is believed that the inferior colliculi may function in reflex responses of turning the head, eyes, and body in response to sounds.

The large cerebral peduncles consist of a dorsal portion called the tegmentum (*tegmentum,* covering) and a ventral portion known as the crus cerebri (*crus,* leg). Most of the fibers in the peduncles are those making up the corticospinal and corticopontine tracts. Dissected out, the peduncles resemble two legs dangling down from the main portion of the brain. The cerebral peduncles are separated from the medial lemnis-

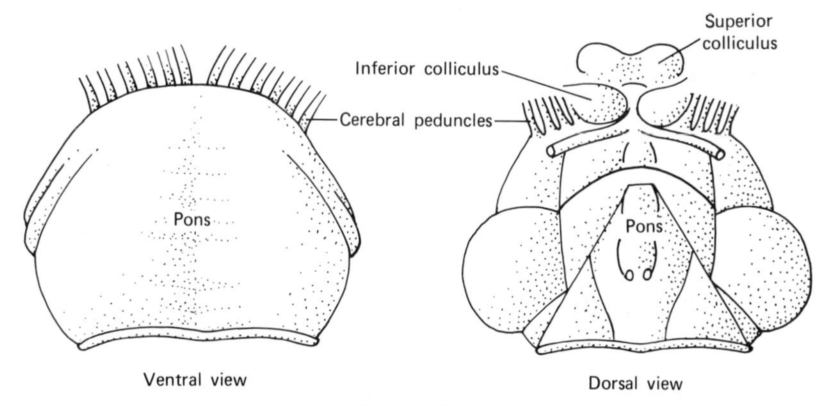

FIGURE 10-7. Ventral and dorsal views of the midbrain in relation to the pons.

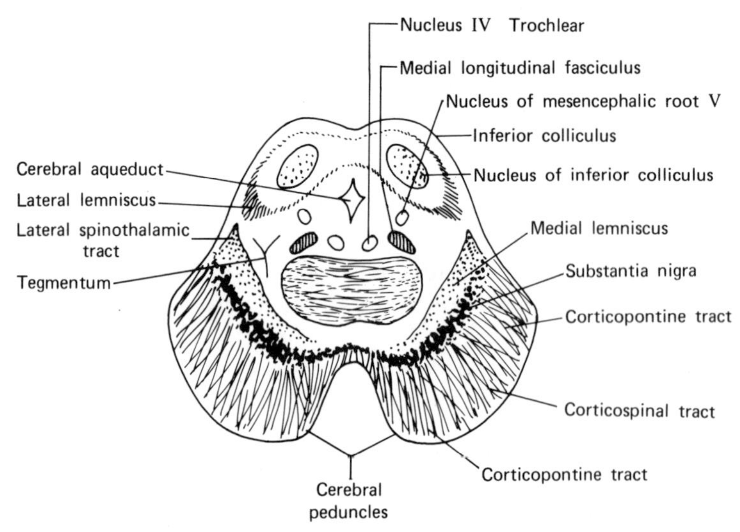

FIGURE 10-8. A cross section of the lower midbrain.

cus by the substantia nigra (*substantis,* substance; *nigra,* black), a dark appearing area of gray matter containing melanin cells. Their function is believed to be that of synthesizing dopamine, a central nervous system excitatory agent.

Two nuclei of the cranial nerves may be identified in this section of the midbrain, the nucleus of the mesencephalic root of the fifth or trigeminal nerve, whose fibers arise from muscle spindles in the facial muscles, and the trochlear nucleus, whose motor fibers supply muscles of the eye. Both nuclei are located close to the midline in the upper third of the midbrain. Several of the more important tracts already discussed in earlier chapters are also identified in Figure 10-8.

A cross section through the more rostral region of the midbrain (Fig. 10-9) reveals the familiar cerebral peduncles on the ventral side with the substantia nigra separating them from the medial lemniscus. One important new pair of nuclei appear in this section, the red nuclei, also commonly known by their Latin name, the nuclei ruber. The red nuclei are part of the extrapyramidal system. They receive input from the dentate nuclei of the cerebellum and send fibers to the reticular formation and to lower spinal levels by way of the rubrospinal tract contributing to the control of muscular activity.

The dorsal portion of the upper region of the midbrain contains the superior colliculi, complex nuclei associated with visual reflexes. Some of the cells of the superior colliculi are involved in the blink reflex, others in

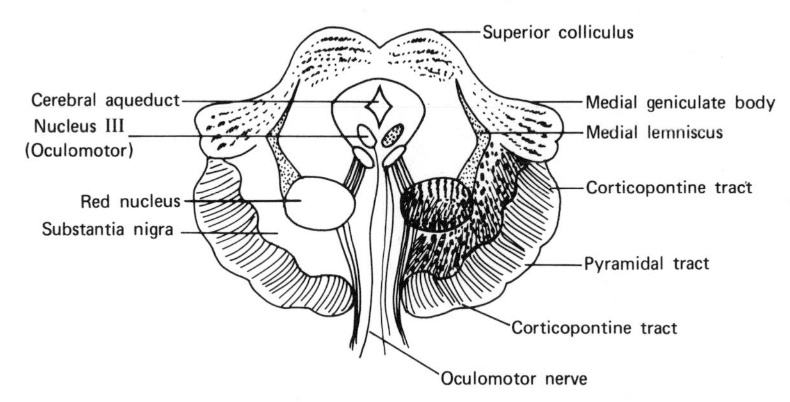

FIGURE 10-9. A cross section of the rostral region of the midbrain. Note the red nucleus appears at this level.

turning the eyes and head in following a moving object, and some in responding to movement in the visual field. Just under the cerebral aqueduct is the oculomotor nucleus, the point of origin for the motor component of the third cranial nerve.

On the upper lateral portion of this region of the midbrain the medial geniculate bodies begin to make their appearance. However, these are associated with thalamic functions and their discussion will be deferred until Chapter 13.

INJURIES TO THE MIDBRAIN: THE BODY LOSES

Lesions to the basal midbrain may involve the oculomotor nuclei, the red nuclei, or the fibers of the pyramidal tract as they course down through the cerebral peduncles. The associated symptoms in the case of a lesion in the region of the oculomotor nuclei (or their fiber pathways) on the right side is lateral strabismus of the right eye, loss of ability to raise the right upper lid, and dilation of the pupil of the right eye. This syndrome is the result of the interruption of the somatic and parasympathetic pathways in the oculomotor nerve.

When one of the red nuclei is involved, there may be tremors in the arm and leg on the same side because of interruption of fibers from the cerebellum to that nucleus.

If one of the cerebral peduncles is involved, there will be spastic paralysis of the opposite arm and leg because of the interruption of the crossed upper motor neurons of the corticospinal tracts.

TABLE 10-1
The Principal Components of the Medulla, Pons, and Midbrain

Division	Prominent Longitudinal Tracts and Functions	Cranial Nerve Nuclei	Other Nuclei	Other Important Structures
Medulla	1. Fasciculi gracilis and cuneatus—kinesthesis to the cerebral cortex	1. Nucleus ambiguus—motor components of the glossopharyngeal, vagus, and spinal accessory nerves	1. Medullary rhythmicity center for the control of respiration	1. Reticular formation alerting the cortex and controlling sleep
	2. Anterior and posterior spinocerebellar tracts—unconscious kinesthesis to the cerebellum	2. Solitary—for afferent fibers of vagus		
	3. Lateral spinothalamic tract—pain and temperature	3. Vestibular—vestibular branch of the auditory nerve		
	4. Pyramidal tracts—motor	4. Hypoglossal—for the hypoglossal nerve		
	5. Rubrospinal tract—vestibular reflexes			
Pons	1. Corticospinal tracts—motor control from cortex	1. Nucleus of the abducens	1. Olivary nuclei—relay centers for auditory fibers to cortex	1. Middle cerebellar peduncle—links cerebellum to pons

	2. Corticopontine—linking cortex to cerebellum	2. Facial nucleus	2. Dentate—connections from cerebellum to basal ganglia	2. Trapezoid body—auditory relay fibers to the olivary complex
	3. Medial lemniscus—fibers from nucleus gracilis and cuneatus	3. Motor for trigeminal	3. Lateral and superior vestibular nuclei—vestibular reflexes	
	4. Lateral spinothalamic tract—pain and temperature	4. Sensory for trigeminal		
	5. Medial longitudinal fasciculi—vestibular reflexes and reticular formation			
Midbrain	1. Medial lemniscus	1. Oculomotor	1. Superior colliculus—optic reflexes	1. Cerebellar peduncle—linking the cerebellum and pons
	2. Medial longitudinal fasciculus	2. Trochlear	2. Inferior colliculus—auditory reflexes	2. Corpora quadrigemina—the superior and inferior colliculi
	3. Corticospinal tract	3. Trigeminal	3. Substantia nigra—links cerebellum to basal ganglia	3. Decussation of cerebral peduncles
	4. Corticopontine tract		4. Red nucleus—connecting midbrain to basal ganglia	

We might also note that lesions to the medial lemniscus result in a loss of tactile, kinesthetic, thermal, and pain sensitivity from the opposite side of the body. This is caused by interruption of sensory fibers ascending from the spinal pathways, particularly those synapsing in the nuclei of gracilis and of cuneatus and those forming the spinothalamic tract.

PUTTING IT ALL TOGETHER

Now that we have navigated the difficult passage from the spinal cord to the cerebrum and cerebellum by way of the medulla, pons, and midbrain, it seems appropriate to attempt to consolidate the components of these structures by tabulating their more important pathways, nuclei, and other structures located in this division of the brain. Table 10-1 summarizes and relates the components that we have identified in this and the preceding chapter.

11 THE RETICULAR ACTIVATING SYSTEM

Ramon y Cajal first described the reticular formation over a half-century ago. His stained sections of the lower division of the brain and upper part of the spinal cord revealed a central core of tangled neural tissue within the ascending and descending sensory and motor tracts. Because cell bodies intermingled with proliferating fibers gave the appearance of a loosely organized network, the structure was named the reticular formation.

From anatomical and physiological studies carried out over the past generation, we know that this complex mass of tissues extends from the spinal cord up through the medulla, pons, and brainstem to the thalamus and then beyond to the cerebral cortex. The cortex, in turn, sends fibers down into the lower levels of the reticular formation, providing for both an ascending and descending network. A rich source of input for the ascending reticular formation comes from the collateral neurons in the medulla and pons that branch off from the ascending sensory tracts in the spinal cord (Fig. 11-1).

In this phylogenetically old and diffuse system of nuclei and tracts, neurophysiologists believe that they have discovered important centers for alertness, sleep, attention, and possibly for that greatest mystery of the nervous system, consciousness itself. Diffuse electrical stimulation applied to the mesencephalic, pontine, or medullary levels of the reticu-

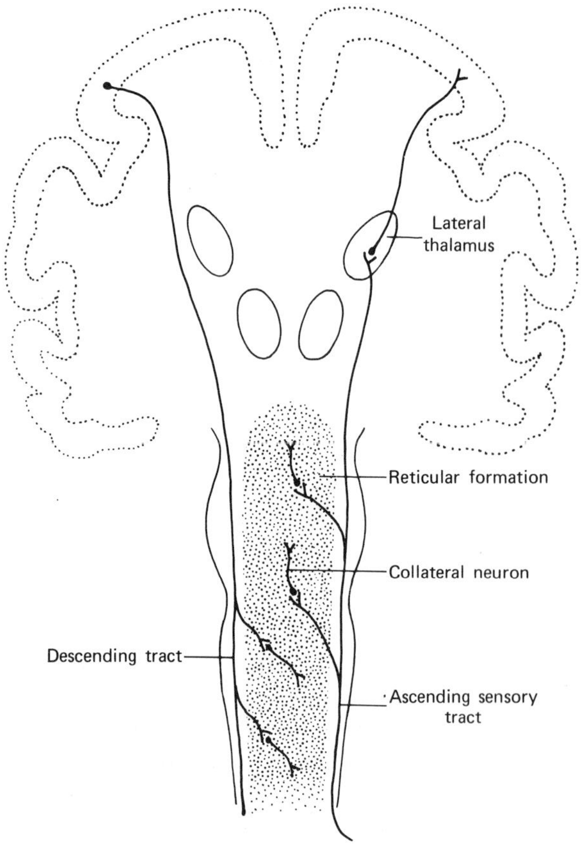

FIGURE 11-1. Collateral ascending and descending tracts providing input for the reticular formation.

lar formation immediately alerts the cerebral cortex, even in a sleeping animal. This effect is called the arousal reaction and may be measured neurologically by electrical changes picked up from the surface of the cerebral cortex.

The state of alertness produced by reticular stimulation persists for a period of time even after the original stimulus is removed. If, on the other hand, large segments of the reticular formation are destroyed, the animal sinks into a comatose state from which it cannot be aroused. A similar condition is frequently observed in human patients who are the victims of disease or of accidents that severely damage the reticular formation. The barbiturate drugs and certain anesthetics have a depressing effect on the reticular formation without simultaneously blocking the surrounding fibers of the specific sensory system. Conversely, drugs such as norepinephrine increase the level of excitation of the reticular formation.

ACTIVATION OF THE CORTEX: THE ELECTRONICS OF CONSCIOUSNESS

The cerebral cortex shows some degree of intrinsic electrical activity over the entire trajectory of life from birth to death. We know of this intrinsic electrical activity from studies made with the electroencephalogram, which reveals signs of cortical activity even when the subject is in a coma. At medical centers throughout the United States, the increasingly accepted clinical test of death is the complete absence of electrical signs of brain activity.

There are several possible sources of cortical activation, one of which is the cortex itself. As we shall find (Chap. 16), the cerebral cortex consists of an outer layer of matted dendrites whose cell bodies lie at deeper layers. It is the summated activity of these cells that the electroencephalogram records. That the tissues of the cortex show intrinsic signs of electrical activity can be demonstrated by recording potentials from freshly excised slices of animal brains. However, there are other sources of activation of the cortex, and the electroencephalogram records the total effect of these sources of stimulation on the network of cortical cells.

The first extrinsic source of cortical stimulation comes from the ascending sensory tracts in the spinal cord and the sensory components of the cranial nerves. Because both of these systems mediate specific sensory processes, and since they are routed through nuclei in the thalamus, they comprise what is called the specific thalamocortical projection system (Fig. 11-2). The nuclei for this system lie in the lateral portion of the thalamus, and it is here that all sensory neurons—except for those for the olfactory system—synapse with the projection fibers conducting sensory impulses to the appropriate areas of the cortex.

The second source of stimulation for the cortex comes from the nonspecific or diffuse thalamocortical system. The nuclei in the thalamus serving this system are not discretely localized as are those for the specific sensory pathways. They lie in the midline region of the thalamus and receive input from the reticular formation (Fig. 11-2). These midline thalamic nuclei send fibers to the cerebral cortex, thus serving as a relay station for the ascending reticular formation. We are now in a position to understand how cortical activation is controlled by these various input systems.

Stimulation of the medial thalamic nuclei at rates of 8–12 per second

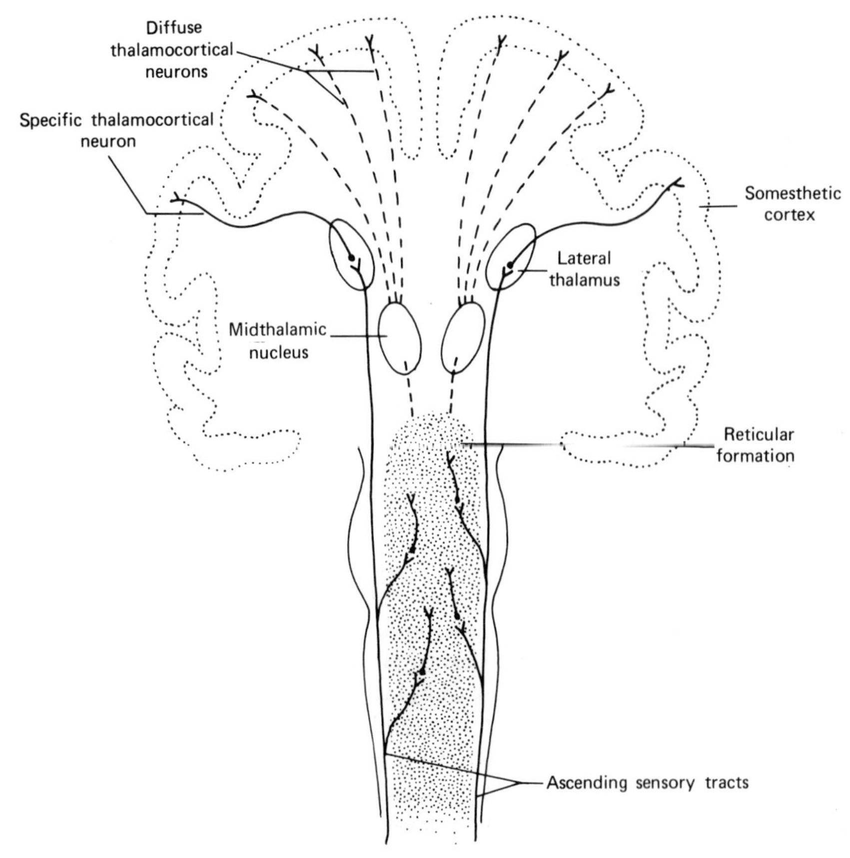

FIGURE 11-2. The specific thalamocortical projection system (solid lines) and the diffuse thalamocortical system (dashed lines).

generates large potential changes on the surface of the cerebral cortex called recruiting waves because at first they are relatively weak but become stronger and stonger with repeated stimulation since additional synapses are being "recruited" into action. Stimulation of the medial nuclei generate waves that are slow in onset but relatively long lasting. These waves appear in widely scattered areas of the cortex. Stimulation of the specific nuclei of the thalamus generates waves that are rapid in onset but relatively short in duration. They are, moreover, localized in restricted areas of the cortex.

These findings suggest that the nonspecific thalamic system, in conjunction with the pontine and medullary reticular systems, controls the overall level of activation of the cortex. Therefore, when incoming signals from the specific sensory impulses to the cortex are received, the cortex is in a state of heightened excitability and ready to receive them.

TO SLEEP, PERCHANCE TO DREAM

The lowest normal level of activation of the brain is seen in deep sleep, the highest in convulsions—a paradox, since both are states of unconsciousness. However, somewhere between these two extremes are the levels of activation characteristic of drowsiness, wakefulness, and keen alertness. Dreaming, too, has its special level of brain activation—a surprisingly high one, in point of fact.

Levels of cortical activation can be recorded by the electroencephalograph, an elaborate machine that is capable of detecting brain waves from electrodes attached to the surface of the skull, amplifying them, and recording them on moving strips of paper. These waves represent electrical potentials of 0–300 microvolts and frequencies of 1–50 per second. Some are irregular, others show distinct patterns. Those that are patterned have been classified into four major groups: alpha, beta, theta, and delta wave patterns. Figure 11-3 shows the characteristics of several types of waves, including their frequency and the typical conditions under which they are found.

Of particular interest are the alpha waves, whose frequencies are 8–12 per second with potentials of approximately 50 microvolts. Alpha waves occur most prominently in the occipital region of the brain when the individual is resting comfortably and is not attending closely to sensory stimuli or problem solving. The alpha rhythm appears during sleep or when the individual's attention is directed to a sensory stimulus or when he or she comes to grips with a mental problem.

Beta waves occur at higher frequencies and tend to appear during high levels of activation of the cerebral cortex.

Theta waves, of slow frequencies and relatively high amplitudes, occur in the parietal and temporal regions, particularly in children with behavior disorders or in adults who are undergoing emotional stress or frustration, a not uncommon state of affairs in our society.

Delta waves with their slow frequencies of less than about 4–5 cycles per second occur in infants, in adults in deep sleep or anesthesia-induced unconsciousness, and also in cases of severe brain damage.

Brain waves are useful to the clinician as diagnostic aids in suspected tumors of the brain and in certain convulsive and behavioral disorders (see Case IX).

Brain wave patterns also show marked changes during sleep. We

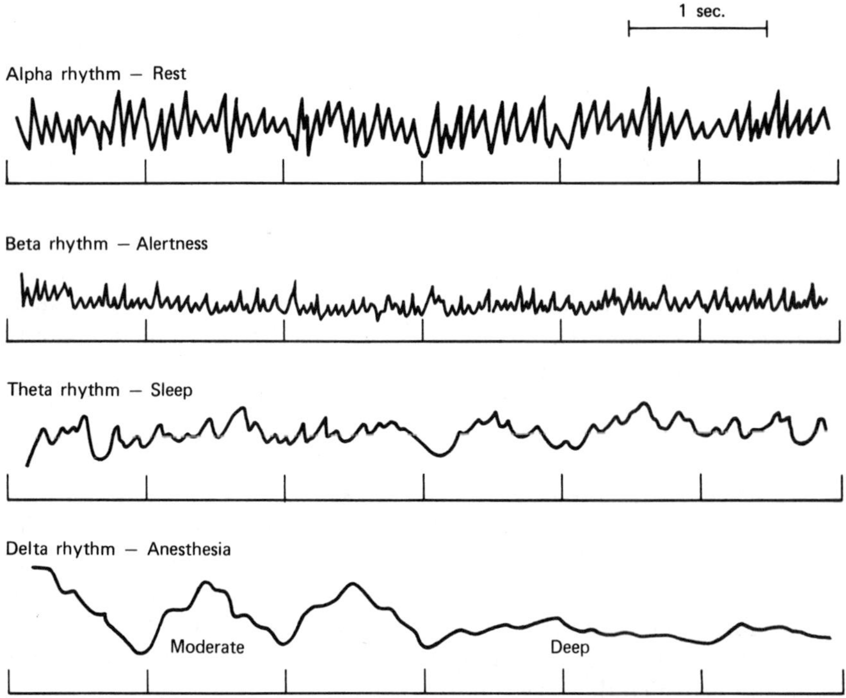

FIGURE 11-3. Alpha, beta, theta, and delta rhythms from the cerebral cortex recorded by means of the EEG.

have already described the slow delta waves characteristic of deep sleep. However, there are two types of sleep, slow-wave sleep and paradoxical sleep. Slow-wave sleep—the kind that knits up the raveled sleeve of care—occurs as a result of decreased reticular activity. We attempt to induce it by relaxation and shutting out sensory stimuli. It is called slow-wave sleep because during this type of sleep the electroencephalogram shows slow delta waves. Paradoxical sleep is characterized by a high degree of brain activation in the form of low voltage beta waves, and yet the individual is *more* difficult to awaken than he or she is during ordinary slow-wave sleep. This is why it is called "paradoxical" sleep. Moreover, paradoxical sleep is associated with rapid movements of the eyes, and so is also known as REM (rapid eye movement) sleep. It was discovered some two decades ago that paradoxical sleep is associated with dreaming and occurs in cycles about every one and a half hours during the night. The duration of REM cycles is 5–30 minutes. Apparently human beings need to engage in paradoxical sleep, for if a person is awakened for several nights every time rapid eye movements occur, he or she becomes restless, tense, and if permitted to return to normal sleep patterns, makes up the deficiency by engaging in longer than average bouts of paradoxical sleep.

Michel Jouvet, a well-known French investigator of sleep, has discovered that midline nuclei in the brainstem in cats, the raphe nuclei, produce serotonin, a hormonelike substance. If these nuclei are destroyed, the animal becomes an insomniac, spending less than 10 percent of its time in sleep as compared to a normal 65–70 percent. Cats (and hedgehogs) are notorious sleepyheads! Another set of nuclei in the pons, known as the locus ceruleus, produces epinephrine. If these nuclei are destroyed, the cat no longer engages in paradoxical sleep. These

CASE IX

THE EEG IN EPILEPSY

C.R., a ten-year-old girl, with a history of convulsive seizures extending back to age two and one-half, was admitted to the medical center for electroencephalography. C.R. has a fraternal twin sister with a similar history who has been successfully controlled by Dilantin.

C.R.'s entire EEG was dominated by seizure activity. The record showed a pattern of bilateral synchronous and symmetrical 2.5–3-second spikes followed by slow waves (see illustration). Occasionally polyspike and slow wave complexes were noted.

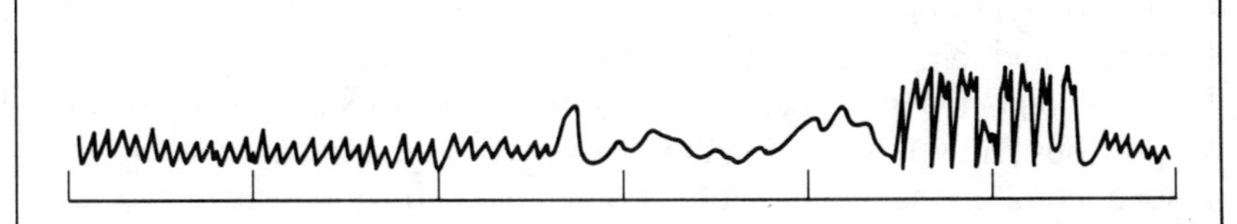

During the examination the patient experienced a 3-second tonic arm extension whereupon the spike-slow wave pattern was replaced by medium amplitude 22 Hz fast waves from all regions of the head.

On the basis of the electroencephalogram and her history, C.G. was diagnosed as a case of idiopathic epilepsy. It was recommended that she be maintained on a regimen of Dilantin and phenobarbital.

CASE X

THE EEG IN BRAIN TUMOR

Mrs. G.B., a forty-two-year-old woman, suffered three grand mal seizures accompanied by left temporal headaches. There was no previous history of convulsive behavior. An EEG examination revealed persistent high amplitude irregularity in the left temporal region with 2–3 Hz delta focus activity. Also noted were irregular low amplitude 4–7 Hz theta waves prominent in the left parietal and temporal regions.

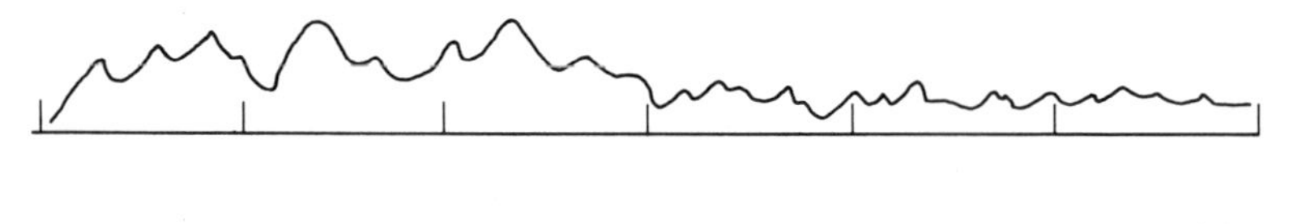

1 sec.

No indication of epileptiform activity was noted. These indices were considered compatible with a destructive lesion in the parietotemporal region. However, x-rays of the patient's skull showed no abnormality, and a CAT (computerized axial tomography) scan revealed no focal lesions. Moreover, there were no visual disturbances, no motor or sensory defects, and no reflex pathology. G.B. was therefore put on Dilantin and phenobarbital and discharged.

Within a month G.B. was readmitted because of a recurrence of seizures and the development of a number of alarming symptoms, including numbness of the right hand, aphasia, agraphia (inability to write), and temporary periods of inability to move her right arm.

A repeated electroencephalogram showed exaggeration of the previous pattern. An angiogram revealed a multivascular area the size of a lemon in the left parietotemporal region above the angular gyrus. A subtotal excision of a glioblastoma was carried out followed by radiotherapy. However, the patient's condition deteriorated, followed by her death within three months.

findings suggest that serotonin ordinarily stimulates light sleep by depressing the reticular activating system, while paradoxical sleep is induced when a pattern of alertness is superimposed upon deep sleep. This pattern of alertness resembles wakefulness except for loss of muscle tension. Unfortunately, felines do not show the deep sleep characteristic of human subjects, and so we cannot generalize the mechanisms of sleep in human beings from research on cats.

There are regions of the brain in experimental animals that will produce sleep when stimulated. As we have seen, there are also areas that will produce wakefulness if removed. Some investigators, therefore, have suggested that there are specific centers for sleep, that neurologically it is an active process and not a passive one, not merely a diminution of neural activity or fatigue. However, fundamentally we do not yet understand completely the physiology of sleep, wakefulness, and consciousness, but as the work of the past several decades we have outlined shows, giant strides toward the solution of these mysteries are being taken.

THE RETICULAR FORMATION AND DECEREBRATE RIGIDITY: A CARICATURE

It is believed that the reticular formation may contribute to the regulation of motor activity by facilitating excitation to the extensor muscles of the body. The role played by the reticular formation in muscular control is most clearly seen in the decerebrate preparation, an animal in which the brainstem is sectioned just anterior to the medulla. This leaves the posterior part of the reticular formation intact in its connections with the spinal cord. The decerebrate animal shows exaggerated contraction of the extensor muscles with the legs, neck, and tail extended rigidly. The spinal animal by contrast is flabby and lacking in muscular tonus. The rigid posture of decerebrate animals suggests that the reticular formation may relay excitatory impulses to the motor neurons controlling the extensor muscles. In the decerebrate animal the effect is exaggerated extensor posture—what Sir Charles Sherrington called a "caricature of reflex standing." In the normal animal excessive extensor excitation is controlled by higher centers, those that are missing in the decerebrate preparation. The exaggerated muscle reaction seen in decerebrate rigidity and in certain types of spastic paralysis exemplify what has been called escape from higher inhibition.

12
THE CEREBELLUM

The ancients, knowing more about structure than function, believed that the cerebellum was a battery of animal spirits for energizing the body. In the same mistaken way, they thought the cerebrum with its richly fissured surface might be a radiator for cooling the blood. It was not until the brilliant work of Pierre Flourens, the French neurologist, experimentally established levels of functioning for various parts of the brain that the proper role of the cerebellum was discovered (Experiment VII). Upon removing successive layers of the cerebellum in pigeons, Flourens found that "the faculty of coordinating movements into walking, jumping, flying, or standing depends exclusively on the cerebellum." Today's clinician, fully aware of the importance of the cerebellum in coordination, observes a child walking with legs apart, dizzy, staggering as if intoxicated, falling, and vomiting explosively, and strongly suspects a cerebellar tumor, a rapidly growing proliferation of cells that may eventually invade the medulla and kill the victim (Case XI).

THE LITTLE BRAIN: MANY WRINKLED FOLDS

About one-eighth the size of the cerebrum, the cerebellum is tucked in below the larger organ just rostral to the medulla and pons. It is connected to the brainstem by three pairs of peduncles containing afferent and efferent tracts and is covered with meninges that fold over the anterior surface to separate it from the cerebrum.

There are two schemes for describing the gross anatomy of the cerebellum. One of these divides it transversely (Fig. 12-1) into three main lobes. First, the flocculonodular lobe (*floccus,* a tuft; *nodus,* a knot) lies ventral to the main body of the cerebellum. Second, the posterior lobe or main portion lies between the flocculonodular lobe and the third division, which is the anterior lobe lying adjacent to the cerebrum.

This older transverse classification recognizes both basic functional and phylogenetic differences in various parts of the cerebellum. The flocculonodular lobe is the first part of the organ to develop embryologically and is therefore called the archicerebellum (*archein,* to begin or lead). It is closely associated with vestibular functions.

The anterior lobe and part of the posterior lobe are classified as paleocerebellum (*paleo,* old), since these regions are prominent in lower forms and probably developed when animals left the sea to live on land. It is functionally associated with the control of the antigravity muscles. In primates the major portion of the posterior lobe is the neocerebellum or phylogenetically newer portion whose connections are primarily with the motor areas of the cerebral cortex for mediating interactions of the two parts of the brain in the control of muscular coordination.

A more recent scheme, and one that is somewhat simpler, utilizes a longitudinal or vertical approach, dividing the cerebellum into three

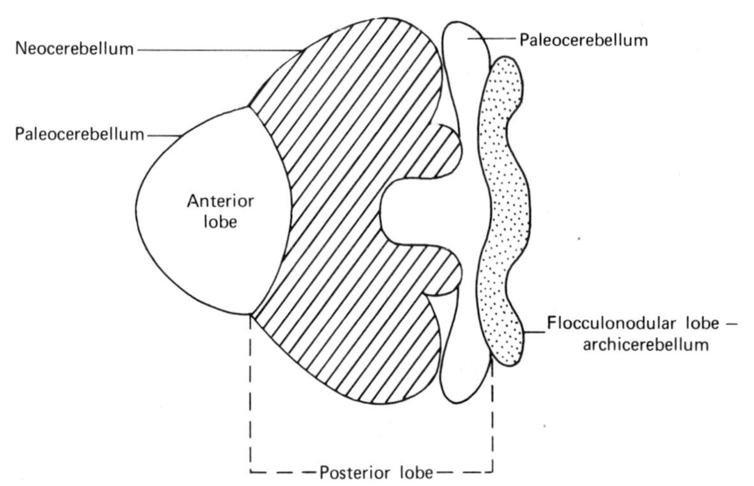

FIGURE 12-1. A schematic diagram of the cerebellum as if it were unrolled showing the three main lobes.

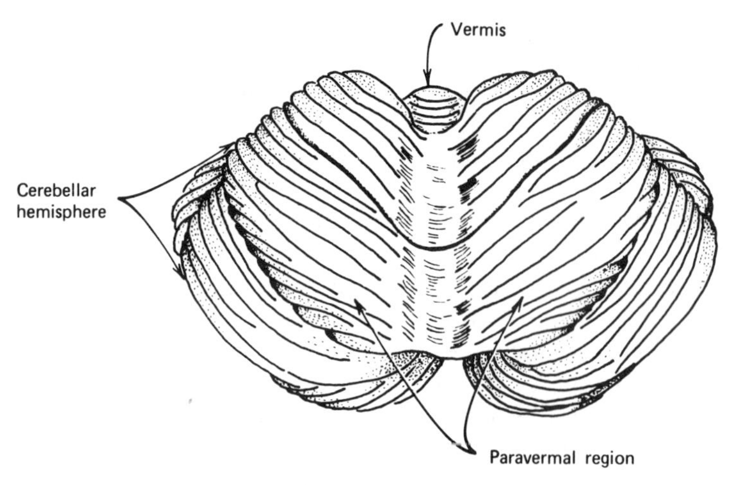

FIGURE 12-2. A dorsal view of the cerebellum showing the main divisions.

main areas (Fig. 12-2). The vermis (*vermis,* worm) is a median region that resembles a worm. On either side of the vermis is the paravermal region, and lateral to this are the two cerebellar hemispheres. The main functions of the median zone are in postural adjustments and movements of the entire body. The paravermal region functions in discrete movements of the extremities as do the lateral regions of the cerebellar hemispheres.

INSIDE: A GARDEN OF MOSSES, CLIMBERS, AND THE TREE OF LIFE

Deep fissures separate the cerebellar cortex into a large number of narrow folia, or folds. Its surface, like the surface of the cerebrum, is made up of gray matter that lies around a core of white matter. Microscopic examination of the cortex reveals three distinct layers, the outer molecular layer, the Purkinje cell (or middle) layer, and the granule cell (or inner) layer (Fig. 12-3). These layers are uniform in structure throughout the cerebellar cortex showing no cytoarchitectural subdivisions in contrast to the cerebral cortex, which varies from one area to another.

The outer molecular layer consists of two types of neurons, dendritic aborizations and numerous axons whose course is parallel to the long axis of the folia. The cells include basket cells and stellate cells. These are

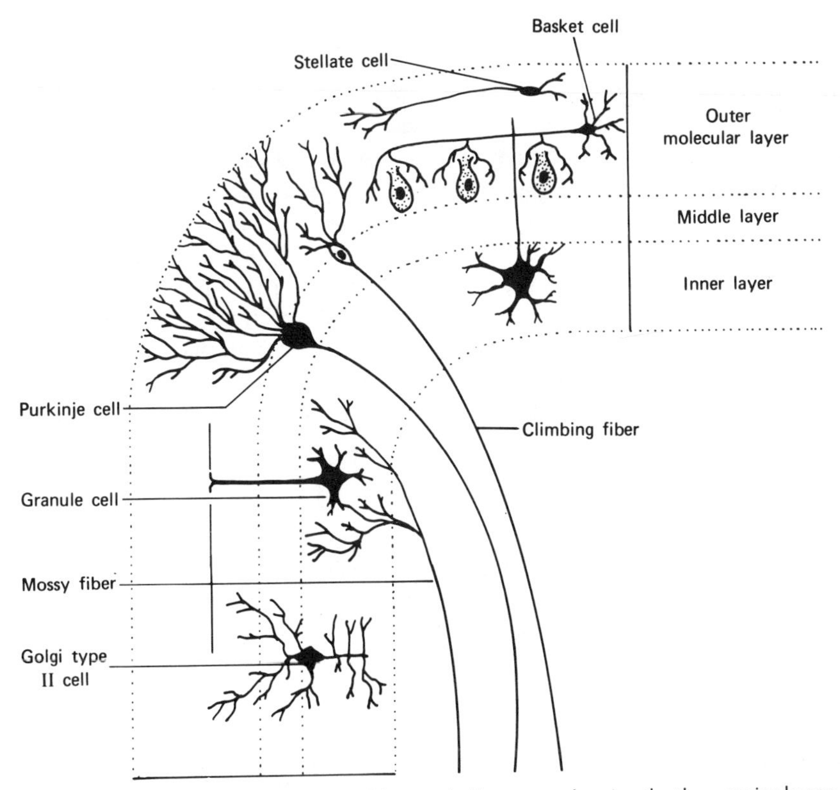

FIGURE 12-3. A schematic diagram of the cerebellar cortex showing the three major layers.

intermixed with dendrites from Purkinje and Golgi type II cells and the transversely oriented axons of granule cells present in the molecular layer.

The middle layer is made up of the conspicuous Purkinje cells with their enormous dendritic processes. It was the resemblance of these cells to the foliage of the white cedar tree that led anatomists to call sections through the cerebellum the arborvita, or tree of life.

The innermost layer is made up of millions of small granule cells and Golgi type II cells. The axons of the granule cells ascend vertically into the molecular layer and bifurcate into branches that run parallel to the long axis of the folia. The Golgi type II cells are less numerous than the granule cells and are located in the uppermost portion of the granular layer.

Synaptic connections in the cerebellum are extremely complex and not yet fully understood. The afferent input arrives over two specialized types of neurons called mossy fibers and climbing fibers (Fig. 12-4). Mossy fibers terminate on large numbers of granule cells, while the climbing fibers terminate on the dendrites of Purkinje cells. Both of these fiber types are excitatory in function. The granule cells, which are also excitatory, can excite basket cells, stellate cells, and Golgi cells. Purkinje cells are inhibitory to the deep cells of the cerebellar nuclei and therefore represent the final output from the cerebellar cortex. Figure 12-4 also shows the excitatory-inhibitory mechanism of the cerebellar cortex and nuclei in highly diagrammatic form. The functional advantage of this arrangement is that it allows for a delicate balance between excitatory and inhibitory impulses that go out to the muscles—the excitatory by way of the corticospinal tracts and the inhibitory by way of

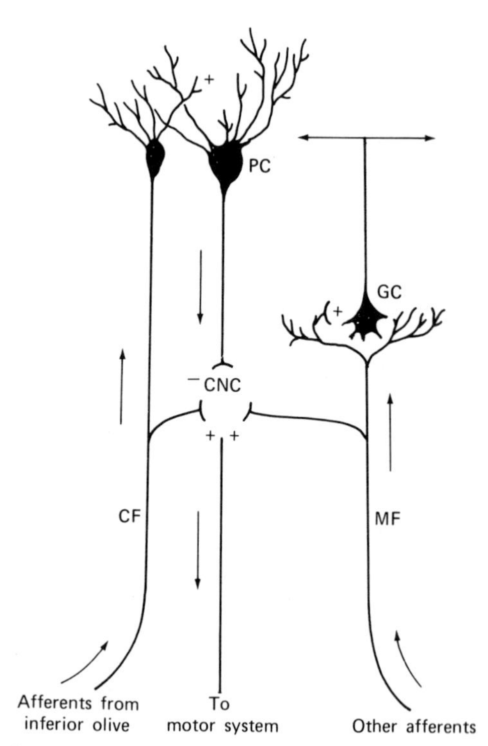

FIGURE 12-4. A diagrammatic representation of the excitatory (+) and inhibitory (−) mechanisms of the cerebellum. CF, Climbing fibers; PC, Purkinje cell; MF, Mossy fiber; GC, Golgi type II cell, CNC, Cerebellar nuclear cells.

tracts originating in the cerebellar nuclei. We shall have more to say about these functions later in this chapter.

CEREBELLAR NUCLEI: TEETH, PEAKS, GLOBES, AND STOPPERS

The most conspicuous nucleus in the cerebellum is appropriately named the dentate for its resemblance to a set of teeth (Fig. 12-5). The other nuclei are the fastigial (*fastigi,* peak or top), located on top of the fourth ventricle; the globose, a globe-shaped nucleus, near the midline; and the emboliform (*embolos,* stopper), an elongated nucleus that lies medial to the dentate. These nuclei are primarily synaptic centers for the efferent fibers leaving the cerebellum.

Entrances and exits to the cerebellum are made by way of three pairs of cerebellar peduncles. They are the inferior, also called the restiform body (*restis,* rope); the middle, also called the brachium pontis; and the superior, also called the brachium conjunctivum (*conjunctus,* joined).

The inferior peduncle on either side consists of a small efferent tract to the vestibular nuclei and a relatively larger number of afferent tracts originating in the vestibular nerve, the olivary nuclei, the spinocerebellar tracts, and the reticular formation.

The middle peduncle is made up primarily of crossed fibers from the pontine nuclei located in the basal pons.

The superior cerebellar peduncle is the primary efferent pathway from the dentate and other cerebellar nuclei.

It should be noted that kinesthetic fibers from the dorsal and ventral

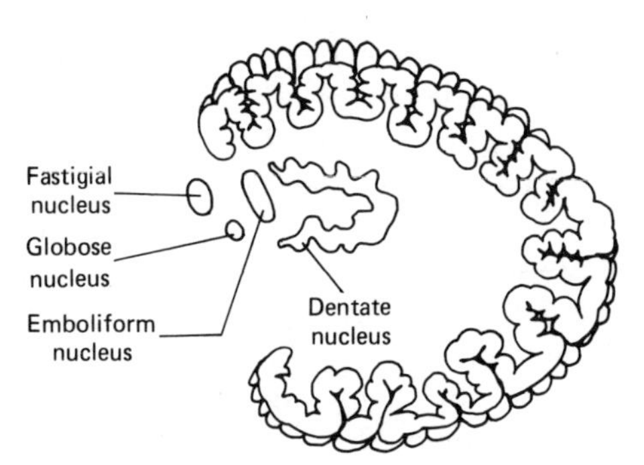

FIGURE 12-5. A saggital section of the cerebellum showing principal nuclei.

spinocerebellar tracts stream up to the cerebellar cortex by way of the superior and inferior cerebellar peduncles.

FUNCTIONS OF THE CEREBELLUM: FINDING THE TARGET

The cerebellum has often been described as a complex high-speed computer that functions as a range finder for the limbs. This analogy assumes that executive commands are sent to the muscles from the cortex culminating in conscious or voluntary motor activities, with smoothing and coordinating of these activities contributed by the cerebellum. This prevents jerky movements or the tendency of the limbs either to overshoot or undershoot their targets. It is precisely this kind of synergy or smoothness and accuracy of movement that the cerebellum supplies by means of its excitatory-inhibitory loops.

Input to the cerebellum about executive signals sent to muscles

EXPERIMENT VII

FLOURENS AND LOCALIZATION OF FUNCTION IN THE BRAIN

Pierre Flourens (1794–1867) was a French experimental neurologist who introduced an important method for the study of the brain—the method of ablation or extirpation in which a part is removed to see what effect it has on the animal's behavior. In developing the method of ablation, Flourens also made an important discovery about how the brain functions, which he reported in 1824 in his landmark book, *Recherches expérimentales sur les propriétés et les functions du système nerveux dans les animaux vertèbres*.

Basically, Flourens proved for the first time that there was localization of function in the brain, but that localization is not so specific as the phrenologists had claimed. Phrenology was a popular early nineteenth-century pseudoscience sponsored by two Germans, Franz Joseph Gall (1758–1828), and his pupil, J.C. Spurzheim (1776–1832), who taught that the brain functions (or "faculties" as they were then called) could be "read" by examining the configuration of the skull. Thus, a high forehead signified strong frontal development of the cerebrum. The frontal regions were identified with intellectual functions. If the area over the cerebellum was prominent, this signified an individual with strong sexual drives. The evidence for phrenology was largely anecdotal or based on the uncontrolled observation of clinical cases.

Utilizing small animals and birds, Flourens attacked the problem of localization of function experimentally. Here is his report on the functions of the cerebellum:

> I have removed by successive layers, the cerebellum of a pigeon. During the ablation of the first layers, there appeared only a little weakness and disharmony in movements. With the removal of the middle layers, an almost general agitation appeared, although there was no sign of convulsion; the animal performed abrupt and disordered movements. With the extirpation of the last layers, the animal whose faculty of jumping, flying, walking, and standing had been more and more disrupted by the preceding mutilations, lost this faculty entirely. From all these facts combined, it follows that the faculty of coordinating movements into walking, jumping, flying, or standing depends exclusively on the cerebellum.

Employing similar methods, Flourens concluded that the cerebrum functions in intelligence, perception, memory, and willing. The midbrain he identified with visual functions and the medulla with the vital functions of respiration and heartbeat.

Flourens's brilliant research technique became a model for generations of neuroscientists, and his conclusions about levels of functioning demolished the pseudoscience of phrenology.

comes from sources in the cerebrum by way of the corticopontine, the olivocerebellar, and the reticulocerebellar tracts, all of which are connected to motor centers in the cerebral cortex (Fig. 12-6). In addition, information from the muscles is fed back directly over the spinocerebellar tracts, and, at the same time, vestibular input arrives from the vestibular system to the flocculonodular lobe. There is also evidence that the various special senses—particularly vision—also have cerebellar input, since the senses contribute significantly to the guidance of muscular activities.

Efferent connections from the cerebellum do not go directly to the muscles; rather control is achieved indirectly by tracts originating in the dentate, globose, and emboliform nuclei, which project to the ventral nuclei of the thalamus, and the red nucleus in the midbrain. These, in turn, have connections with the motor cortex. Because of the manner in which the efferent tracts cross, cerebellar control is ipsilateral or on the same side (Fig. 12-7). For this reason injuries to the right cerebellum

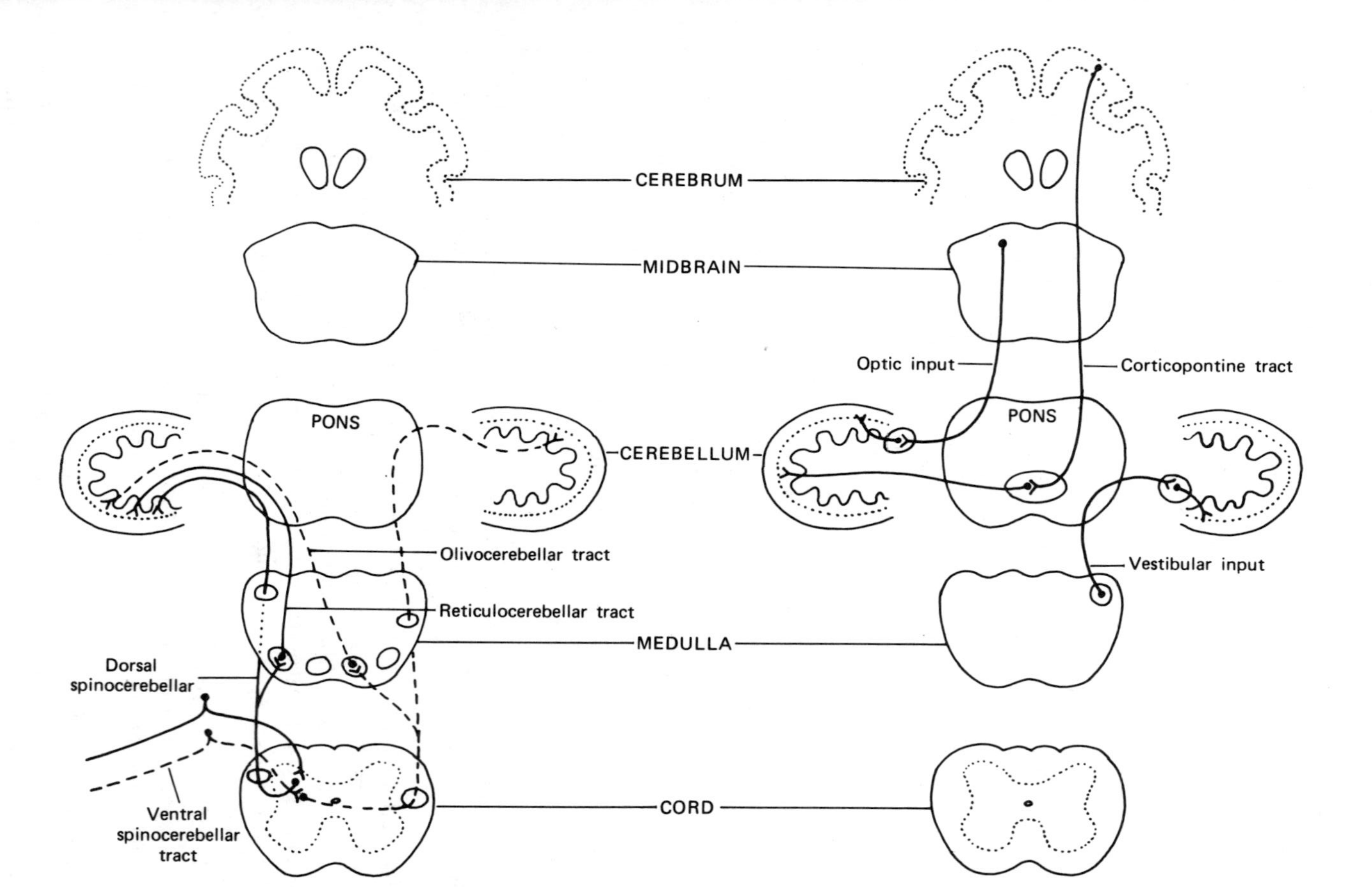

FIGURE 12-6. Afferent input to the cerebellum from the ventral and dorsal spinocerebellar tracts, the medulla, midbrain, and cerebral cortex.

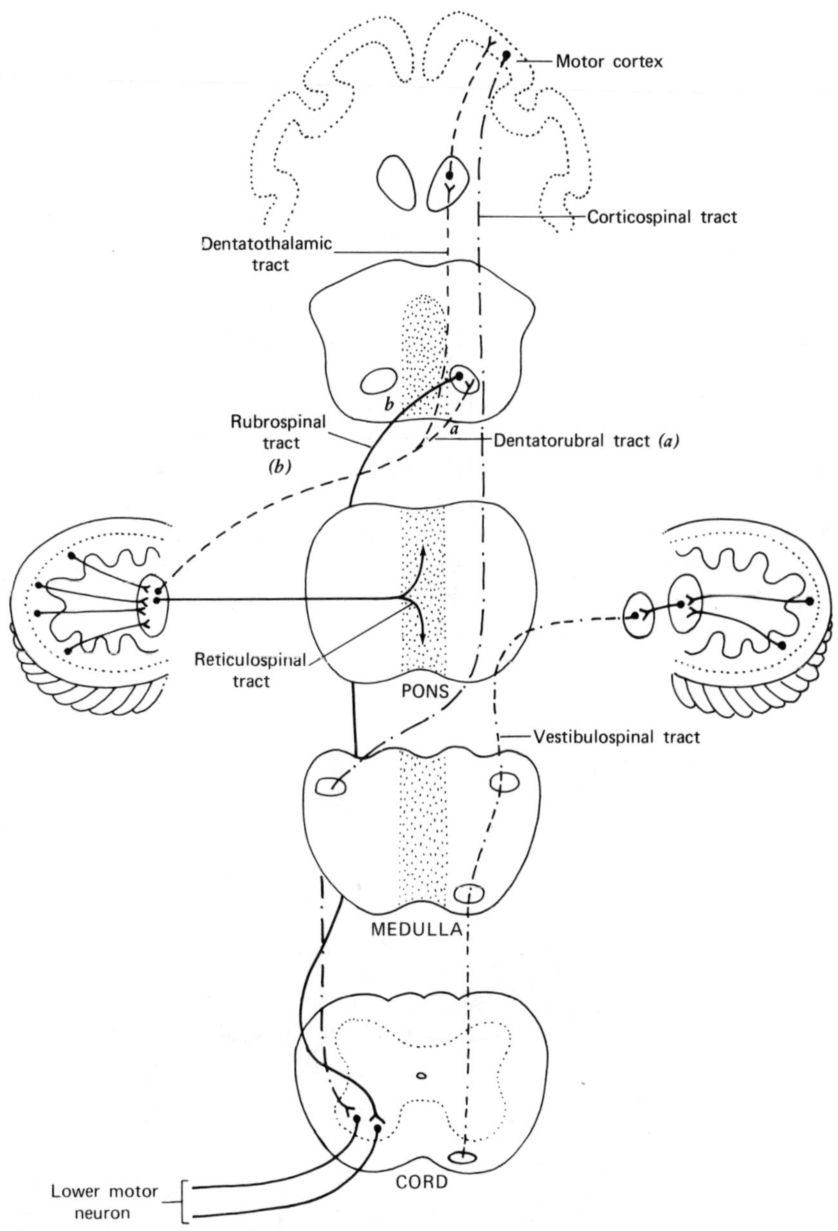

FIGURE 12-7. Efferent output from the cerebellar nuclei. The lateral corticospinal tract is included to show its relationship to the cerebellar system.

affect coordination on the right side of the body, and injury to the left cerebellum affects muscular systems on the left side of the body.

There are also projections from the flocculonodular lobe or archicerebellum by way of the fastigial nucleus to the vestibular and reticular systems for cerebellar control of static and equilibratory functions.

Studies making use of the evoked potential technique show that there is topographic localization in the cerebellum. This is shown in Figure 12-8, where the anterior lobe shows a reversed organization with the hindlegs represented near the front of the cerebellum and the forelegs and head toward the rear. In the posterior lobe, the relationship is just the opposite, with the head being represented anteriorly and the legs posteriorly.

In summary, the cerebellum is analogous to a complex servomechanism that receives and processes information from many sources. It does not initiate movement, but dampens any sudden movement or displacement of the limbs. In a sense, then, the cerebellum acts as an inhibitory agent, smoothing and controlling muscular activity without voluntary or conscious effort on the part of the individual. The importance of these functions is best seen in instances of lesions or diseases to the cerebellum, which we shall consider in the section to follow.

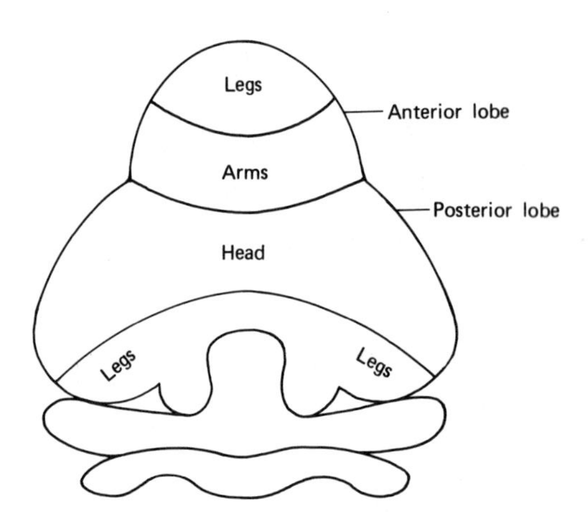

FIGURE 12-8. Topographic localization in the cerebellum.

DISORDERS OF THE CEREBELLUM: MISSING THE TARGET

One of the primary symptoms associated with cerebellar disorders is ataxia, which is characterized by disruption of coordination and timing of movements. The patient walks with a stumbling, falling gait similar to that observed in intoxication. The limbs fail to stop on target so that when asked to place the forefinger on the tip of the nose—a standard

CASE XI

CARCINOMA OF THE CEREBELLUM

C.T., a fifty-three-year-old woman, developed malignancy of the breast necessitating mastectomy. Three years later a diagnosis of metastasis of the malignancy to her cerebellum was made. At that time and again three years later, cerebellar tumors were removed with relief of symptoms.

At the time of the present admission, Mrs. C.T. complained of nausea with vomiting and dizziness, both of which were exaggerated by any attempt at motion. The patient walked as if intoxicated, experienced extreme fatigue, and complained of tenderness in the occipital region of her head.

Diagnostic tests revealed that her sensory and pyramidal systems were intact, but that because of her poor gait, impaired balance, nausea, and past history of cerebellar involvement, a recurrence of the cerebellar tumor had evidently occurred.

In surgery, exposure of the dura and dissection down the midline of the left cerebellar hemisphere revealed thinning and absence of cortical tissue. Palpation of the area resulted in the exuding of an ounce of old blood from an unsuspected hematoma. Further dissection revealed a densely scarred vermis adherent to surrounding tissue and a nodule identified histologically as an adenocarcinoma.

Removal of the scar tissue and nodule with a postoperative regimen of radiation and chemotherapy for a five-month period allowed Mrs. C.T. to be discharged with ability to feed herself, tolerate motion without nausea, and with fair muscular coordination of the lower extremities.

In this patient's case, surgical treatments were palliative rather than curative, and because of the metastasizing malignancy, her prognosis was poor.

clinical test—the patient places it to one side. If the muscles of the larynx are involved, scanning speech may be present, characterized by irregular and incorrectly placed pauses in speaking with particular difficulty in enunciation at the beginning of words and phrases.

Another frequently observed symptom of cerebellar disorders is hypotonia, or decreased muscular tone accompanied by weak tendon reflexes. The clinician tests for hypotonia by simple palpation of the muscles and for reflexes by eliciting the knee jerk with his rubber hammer. If hypotonia is present, the knee jerk is pendular, that is, instead of stopping, the lower limb swings back and forth like a pendulum following the kick response.

Asthenia (*astheneia,* weakness) or muscular weakness is also typically present on the affected side, as is intention tremor. Intention tremor is present when the individual engages in purposeful movements, but not during rest. The affected limb shakes and moves in a coarse, irregular manner, especially near the end of an intended movement.

If the flocculonodular lobe is involved, the patient may be unable to maintain balance and station, having to support himself against objects or walls in order to prevent falling.

Many of these effects are ipsilateral because of the arrangement in the crossing of cerebellar fibers. Moreover, many of the defects in coordination associated with cerebellar disorders are not observed in minor injuries to the organ and may disappear in time (except in malignant tumors) even with fairly severe lesions. The ability of the individual to function well with small lesions and to recover from severe lesions is the result of utilizing compensatory mechanisms—for example, using the eyes for visual guidance, or signals from alternate pathways in the nervous system for maintenance of balance and coordination.

13
THE THALAMUS

In the great mushrooming of tissue that takes place during the embryonic development of the forebrain, a large bilateral mass of gray matter gives rise to a pair of wall-like structures that make up a kind of chamber deep within the cerebrum. This mass is called the thalamus (*thalamos,* bedchamber). It is not hollow, however; rather the thalamus and the rest of the tissues of the diencephalon and telencephalon are fused together, nuclei and fiber tracts all intermixed. To try to understand the whole, the anatomist separates it into parts employing whatever landmarks are at hand in an attempt to make the complex simpler, knowing that parts that are anatomically separable usually have distinct functions. The thalamus has yielded many subparts to the anatomist's knife over the centuries, and modern neuroscientists using electrical methods of stimulation and recording have made great progress in recent decades in assigning functions to those parts. As we explore this hidden chamber of the forebrain, we shall summarize the contribution of this complex nuclear mass to the rest of the nervous system.

A CURTAINED BEDCHAMBER

Named after an old-fashioned bed enclosed by draped curtains, the thalamus is a large mass of gray matter divided by a vertical sheet or curtain of myelinated fibers called the internal medullary lamina (*lamina,* a sheet or thin plate) into three subdivisions, the lateral, medial, and anterior thalamic nuclei. Each of these nuclear clusters can be further subdivided into over two dozen smaller nuclear masses. We can avoid so bewildering an anatomical array by considering the nuclei from a functional point of view and group them into a few basic types.

SENSORY RELAY NUCLEI: MESSAGES FOR HEADQUARTERS

We have already had occasion to note that the sensory fibers ascending the spinal cord terminate in the lateral region of the thalamus. With the exception of the olfactory tracts, all sensory pathways terminate in the thalamus. Therefore, one important function of the thalamus is that of a general relay station for sensory information on its way to the cortex.

As Figure 13-1 indicates, the nucleus ventralis posterolateralis (VPL) and the nucleus ventralis posteromedialis (VPM), which make up the posterior and ventral portion of the thalamic mass, are the sensory relay nuclei whose fibers project to the area of the cerebral cortex concerned

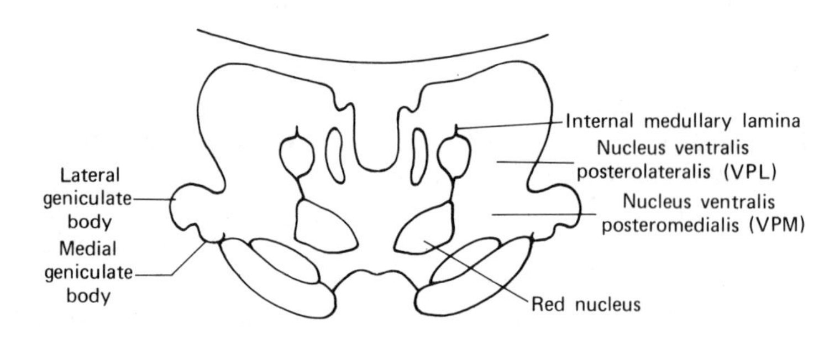

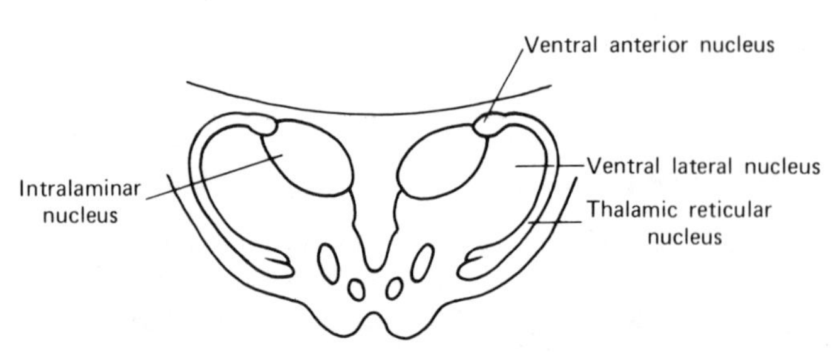

FIGURE 13-1. The major nuclei of the thalamus.

with body sensation or somesthesis, including the special sense of taste.

The medial geniculate body is part of the auditory system, and the lateral geniculate body is part of the visual system. Both of these are also specific sensory relay nuclei for the auditory and visual senses. The medial geniculate body receives fibers from the auditory nerve and projects its own output to the temporal lobe. The lateral geniculate body receives input from the retina and projects to the occipital region of the cerebral cortex (Chap. 7).

It should be noted that all of the sensory relay nuclei receive fibers from the cerebral cortex so that thalamic and cortical functions are closely interconnected with constant interaction back and forth between the two systems.

Returning to Figure 13-1, the ventral lateral nucleus and the ventral anterior nucleus, often collectively called the nucleus ventralis anterolateralis, are motor relay nuclei that receive fibers from the cerebellum and from the globus pallidus (named for its whitish appearance), a center in the forebrain concerned with motor functions that will be discussed in Chapter 15. These relay nuclei project to the motor regions of the cerebral cortex.

The association nuclei make up a large mass of tissue whose fibers project to the association areas of the cerebral cortex. These cortical association areas, to be discussed in detail in Chapter 16, make up the larger portion of the human cerebral cortex and are concerned with memory and integrative functions. As one example useful for our immediate purposes, visual memories are mediated by the visual association cortex, which makes up a large portion of the occipital lobe. Similarly, the lateral posterior, lateral dorsal, and medial dorsal nuclei project to the other major association areas of the cerebral cortex. The functions of the complex interconnections between the thalamic nuclei and the association areas are not yet completely understood. Such a system of connections suggests a close correlation between the association areas of the cerebral cortex and thalamic centers, which enables the brain to integrate incoming sensory information with already stored information. Such an arrangement seems highly adaptive.

OTHER THALAMIC NUCLEI: TO LIMBO AND BEYOND

The part of the human cerebral cortex hidden between the deep longitudinal fissure that almost divides the cerebrum into two halves is known as the cingulate gyrus (*cingulum,* a band or girdle; *gyros,* a circular or spiral form), and is associated with a phylogenetically older part of the brain known as the limbic (*limbus,* border) system. We shall be considering this system in detail in the chapter to follow. However, we may note in passing that the anterior nucleus of the thalamus receives fibers from the hypothalamus and projects fibers to the cingulate gyrus. This portion of the thalamus is, in effect, part of the limbic system.

A large portion of the fibers from the thalamus make up the nonspecific thalamic system, which is not concerned with relaying messages. The fibers of this system originate in the nonspecific thalamic nuclei. Several of these, the intralaminar nuclei, located in the central portion of the mass of the thalamus, along with the reticular nucleus have connections with the cerebral cortex and are the thalamic portion of the reticular activating system whose functions we have already discussed in Chapter 11 (see Fig. 13-1).

INJURIES TO THE THALAMUS: STRANGE SENSATIONS

Lesions to the thalamic relay nuclei, called the thalamic syndrome, typically lead to some loss in body sensatiion on the opposite side. However, sensation to intense stimulation may be present. In some individuals ordinary sensory stimuli are reported as exaggerated or painful. Sensation from the face may be intact, apparently because of alternate pathways in the loosely organized trigeminal-thalamic pathway (see Case XII).

14
THE HYPOTHALAMUS AND THE LIMBIC SYSTEM

It weighs only one-three hundredth of the total weight of the brain, and yet it serves an astonishing number of regulatory functions—those of body temperature, food and water intake, hormonal secretion from the pituitary gland—and influences emotional behavior as well. This small but vital center known as the hypothalamus is not so much a clearly defined structure as a collection of nuclei in the diencephalon with widespread connections to other parts of the brain. There are tracts leading outward to the pituitary gland, the thalamus, the cerebral cortex, the reticular formation, and a part of the brain called the limbic system, which like the hypothalamus is concerned with the regulation of visceral and emotional processes. Because the hypothalamus and limbic system are so closely related, we have chosen to take them up together. In fact, some neuroscientists classify the hypothalamus as part of the limbic system, pointing out how the two work together to program the sequence of activities that satisfy motivational states and regulate those visceral activities associated with homeostasis. It will be convenient to describe first the structure and functions of the hypothalamus and second those of the limbic system.

HYPOTHALAMUS: BETWEEN EYES AND BREASTS

The hypothalamus lies approximately in the middle bottom of the brain (Fig. 14-1). It is situated between the optic chiasma, which marks its rostral limit, and the mammillary bodies (*mamma,* breast), which are its caudal boundary. Internally it forms the floor and lower portion of the walls of the third ventricle. The hypothalamus has been divided into three anatomical regions using as landmarks the optic chiasma, mammillary bodies, and the area between known as the tuber cinereum (*tuber,* bump, swelling; *cinerous,* ashen), or tuberal region.

The supraoptic region is the most rostral lying above the optic tracts; the mammillary region is the caudal portion. Intervening between these

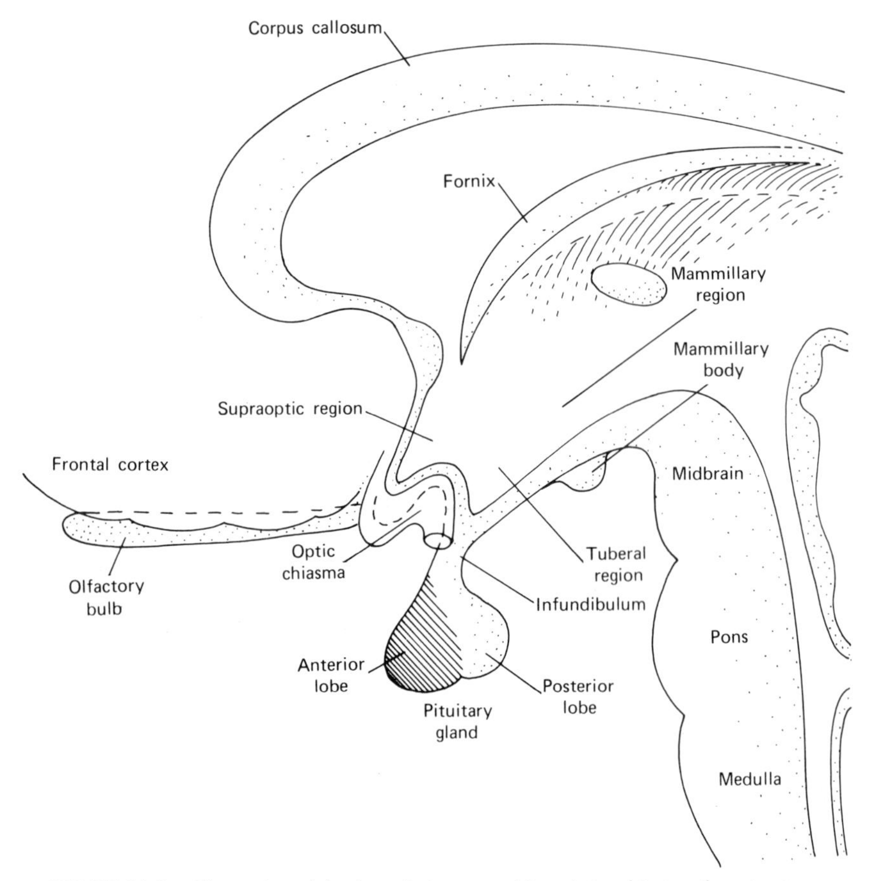

FIGURE 14-1. The region of the hypothalamus and its relationship to other structures.

two is the tuberal region. The most prominent structure in the rostral hypothalamus is the infundibulum (*infundiere,* to pour into as a funnel) or stalk that connects the hypothalamus to the hypophysis (*hypophein,* to grow beneath), or pituitary gland.

NUCLEI AND TRACTS: MANY AND VARIED

The nuclei of the anterior or rostral region are the preoptic, supraoptic, and paraventricular, all named for their spatial relationships to their anatomical landmarks. The tuberal group includes the ventromedial, the dorsomedial, and the arcuate nuclei. Figure 14-2 shows the more prominent nuclei in relation to adjacent brain structures and the pituitary gland.

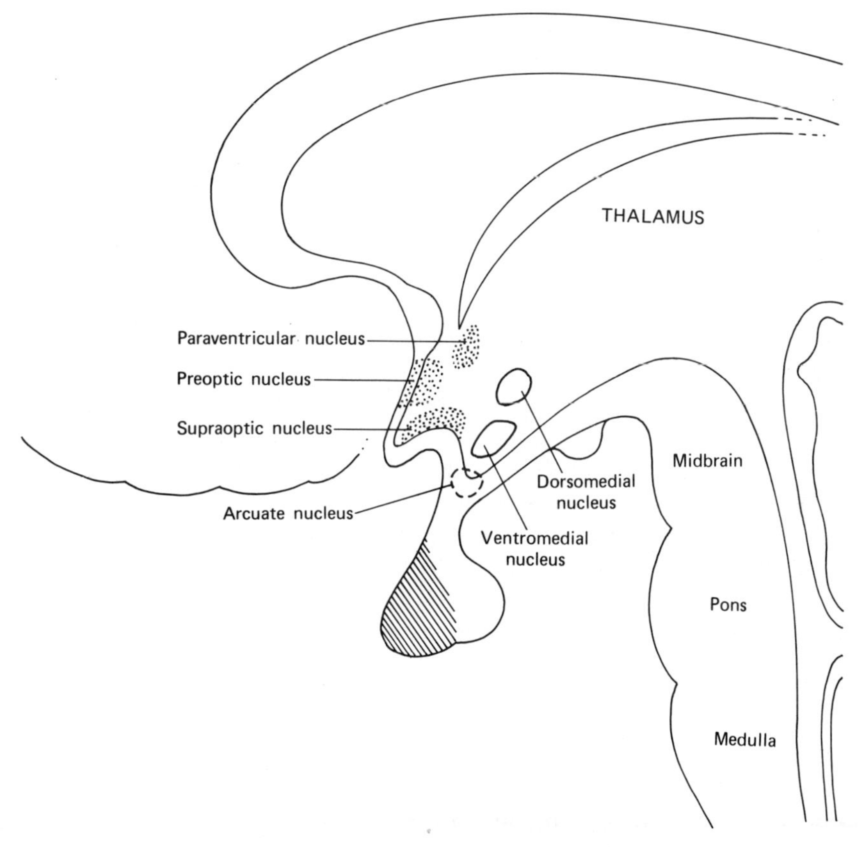

FIGURE 14-2. The nuclei of the hypothalamus.

As Figure 14-3 indicates, these nuclei are connected to other brain structures by both incoming and outgoing fiber tracts. The major afferent or incoming tracts are the following: (1) the olfactohypothalamic, which connects the olfactory area to the hypothalamus; (2) the corticomammillary, which originates in the hippocampus (*hippo,* horse; *kampos,* sea monster), one of the limbic centers to be discussed later, and terminates in the mammillary region; (3) the stria terminalis (*stria,* furrow; *terminus,* end, limit), a fiber tract that extends from the amygdaloid (*amygdala,* almond) complex, another limbic center, to the ventromedial nucleus of the hypothalamus.

The most important efferent projections are (1) the hypothalamohypophyseal tract originating in the supraoptic and paraventricular nuclei and projecting to the pituitary gland (2) the mammillothalamic fasciculus, a tract that originates in the mammillary body and projects to the anterior thalamus; (3) the mammillotegmental fasciculus, a bundle of fibers that projects to the tegmental reticular formation; (4) the preventricular fibers arising in the supraoptic and tuberal nuclei and terminating in the dorsomedial thalamus and various parasympathetic centers.

HYPOTHALAMUS AND HEAT REGULATION: CHILLS AND FEVER

If the brain is transected below the level of the hypothalamus, the animal becomes poikilothermic (*poikilos,* varied) or tends to take on the temperature of the environment in which it happens to be. In the intact animal the temperature-regulating functions of the hypothalamus can be studied more precisely by placing lesions in various parts of the region. If the rostral portion is destroyed, the animal becomes incapable of temperature regulation in a warm environment, although it can continue to regulate its temperature in a cold environment. The deficiency in regulation is due to the failure of heat loss mechanisms, primarily vasodilatation and sweating.

If lesions are placed in the caudal region, the animal fails to maintain heat regulation in either a warm or cold environment and becomes poikilothermic. In this case we have a failure of the heat conservation and production mechanisms—vasoconstriction, shivering, piloerection, and the secretion of epinephrine. It is believed that in addition to destroying the heat conservation center, lesions in the caudal region also interrupt fibers from the rostral regulatory center. In the normal animal the two centers work as a complementary team to maintain temperature

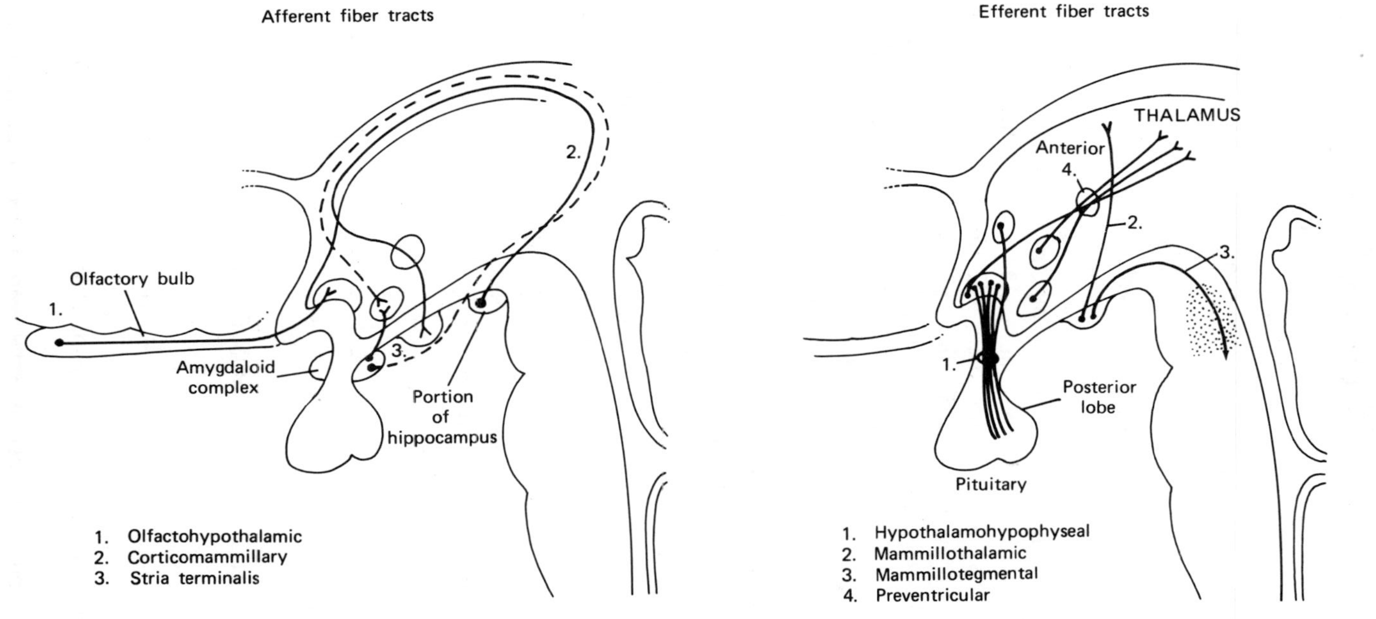

FIGURE 14-3. The afferent and efferent fiber tracts associated with the hypothalamus.

regulation. This interpretation is supported by experiments involving electrical stimulation of the rostral center, which produces panting and vasodilatation, while stimulation of the caudal center causes shivering.

HYPOTHALAMUS AND WATER BALANCE: FLOODGATES AND SPONGES

It will be recalled that fibers from the supraoptic and paraventricular nuclei form the hypothalamohypophyseal tract. It is through this connection that the hypothalamus regulates the production of posterior pituitary hormones, the most important of which is the antidiuretic (*diourein,* to urinate) hormone or ADH, which causes water reabsorption in the kidney tubules. The dramatic effect of a failure of water reabsorption is seen in cases of diabetes insipidus, a disorder caused by destruction of the posterior pituitary gland or by any lesion between the supraoptic and paraventricular nuclei and the posterior pituitary gland. If not maintained on ADH, the victim may lose 20 liters of urine every day, and in an attempt to maintain water balance spend most of the time shuttling between the faucet and the commode.

Water intake as well as output is also under the regulation of the hypothalamus. Experiments on animals suggest that intake is regulated by osmoreceptors in the supraoptic region that are sensitive to the salinity of the blood. Injections of distilled water near the supraoptic nuclei cause immediate cessation of drinking in thirsty animals that have been deprived of water for extended periods of time, while injections of hypertonic solutions of saline initiate drinking in animals that have had unlimited access to water.

EXPERIMENT VIII

HETHERINGTON AND RANSON DISCOVER THE SATIETY CENTER

Walter Cannon, the famous Harvard physiologist, offered a simple explanation of hunger as dependent on "hunger pangs"—periodic and intense stomach contractions stimulated hormonally by the need for food. Experiments in the early part of the twentieth century carried out by Cannon and confirmed by Anton Carlson, a prominent University of Chicago physiologist, provided initial support for Cannon's theory. In these experiments subjects swallowed a small balloon with an attached tube. The balloon could then be inflated and pressure changes in the stomach caused by contractions recorded on a moving strip of paper. These changes seemed to correspond well with subjective reports of hunger.

However, studies of animals whose stomachs had been removed experimentally and of people who had undergone gastrectomies because of disease showed that normal hunger and regulation of food intake can occur in the absence of the stomach.

Recent research on the hypothalamic regulation of hunger experienced a significant advance when A.W. Hetherington and S.W. Ranson, using the Horsely-Clarke technique (a special frame fitted around the head of the animal to guide small electrodes to centers deep within the brain), were able to demonstrate that small bilateral lesions of the ventral hypothalamus cause obesity. Postoperative animals began eating voraciously and increased their weight several times over normal levels.

Subsequent discoveries by other investigators pinpointed an area in the lateral hypothalamus which, when destroyed, leads to anorexia, or loss of appetite, in experimental animals. These findings led to the generalization that the ventromedial hypothalamus is a satiety center and the lateral region a feeding center. Research over the past decade has shown that the regulation of food and water intake is a complex process dependent on a number of physiological and psychological factors. However, the pioneer work of Hetherington and Ranson with microelectrode techniques provided a major technical breakthrough for neuroscientists investigating the role of the hypothalamus in drive regulation.

From A.W. Hetherington and S.W. Ranson, *Anatomical Records,* Vol. 78, 1940, pp. 149–172.

HYPOTHALAMUS AND FOOD INTAKE: WEIGHT WATCHERS

Bilateral lesions of the ventromedial hypothalamic nuclei produce hyperphagia, or overeating, with extreme obesity resulting. When the lesions are placed in the lateral nuclei of the hypothalamus, the result is anorexia, or loss of appetite, and eventual starvation. This experimental finding suggests that the ventromedial nuclei are “STOP” or satiety centers and the lateral nuclei “GO” or feeding centers.

In the normal individual, food intake is, of course, regulated by hypothalamic centers but also by blood glucose level, the fullness or emptiness of the stomach, and psychological factors such as the sight and smell of food, personality makeup, and dietary habits, good and bad.

HYPOTHALAMUS AND ENDOCRINE FUNCTIONS: ANOTHER MASTER GLAND

The hypothalamus is not only a neural center but an endocrine gland as well. It is able to synthesize oxytocin, which causes contractions of the uterus at term along with ejection of milk from the mammary glands. It also synthesizes releasers, which stimulate the anterior pituitary to produce its various tropic hormones or those directed at other endocrine glands. The mechanism by means of which releasing factors from the hypothalamus reach their target in the pituitary gland is through a capillary portal system connecting the hypothalamus with the anterior portion of the pituitary. The releasing and inhibiting functions of the hypothalamus in controlling the pituitary gland and, indirectly, other endocrine glands may be summarized as follows:

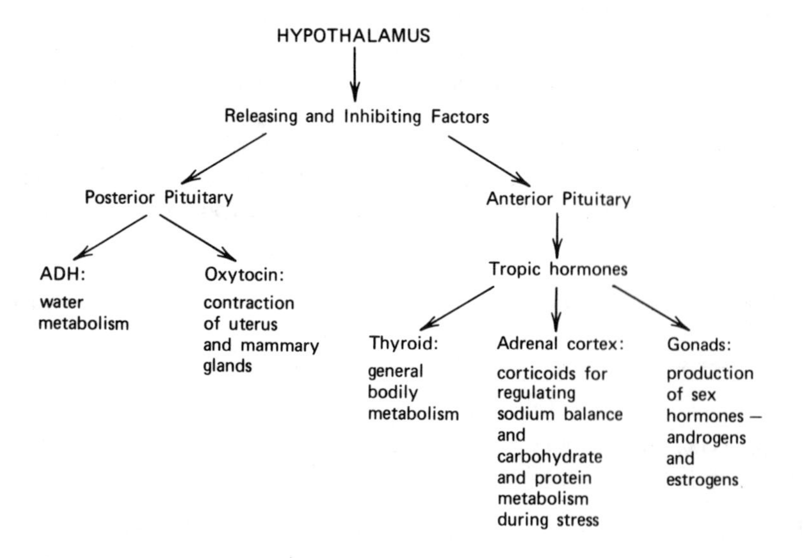

HYPOTHALAMUS AND EMOTIONS: A GENERAL HEADQUARTERS

As we pointed out in Chapter 5, the hypothalamus exerts a general control over the autonomic nervous system. Stimulation of various areas of the hypothalamus can produce most of the effects of sympathetic and parasympathetic activity. Stimulation of the anterior region excites parasympathetic activity while stimulation of the posterior region excites sympathetic effects. Moreover, the integrity of the hypothalamus is necessary for integrated emotional behavior. Animals lacking the hypothalamus show only fragmentary emotional responses and appear incapable of integrated reactions of fear or anger. It is also important to note that the cerebral cortex exerts a restraining influence on the autonomic centers in the hypothalamus. As we pointed out in Chapter 5, the animal lacking a cortex shows "sham rage"—a kind of excessive response generated by the hypothalamus to a mildly painful stimulus. Normally the reaction would be held in check by the cerebral cortex but in its absence assumes an exaggerated pattern.

THE LIMBIC SYSTEM: PLEASURE CENTERS

Figure 14-4 shows the principal parts of what is known as the limbic system. Basically it consists of two divisions: cortical and subcortical. The cortical areas are the cingulate gyrus, which forms a kind of belt around the brain deep within the longitudinal fissure, the hippocampal gyrus, and the pyriform (*peri,* pear), a roughly pear-shaped lobe near the olfactory bulbs.

The subcortical portions of the limbic system are the fornix (*fornix,* arch) an arch-shaped nucleus near the rostral hippocampus, and the septal region in the frontal part of the brain near the roots of the olfactory bulbs. As we have already pointed out, some neuroscientists include the hypothalamic complex in the limbic system.

It has been suggested that sensory or afferent impulses streaming up to the thalamus are funneled through the hippocampus to the various cortical and subcortical centers of the limbic system where they receive their emotional tone. According to some theorists the system is associated with individual and species preservation, with the frontotemporal structures acting in self-protection—fear and anger—in running down and killing the prey, and in sexual responses. Lesions in the amygdala render formerly aggressive and dominant animals docile. In contrast, sexual behavior becomes intensified with autoerotic, homosexual, and heterosexual behavior displayed indiscriminately.

We also know from brain implantation experiments that the septal region contains what appear to be "pleasure centers." Animals will press bars to the point of exhaustion in order to obtain mild stimulating shocks to this region. Some confirmation of the pleasurable nature of stimulation in this region has been obtained from terminally ill patients who

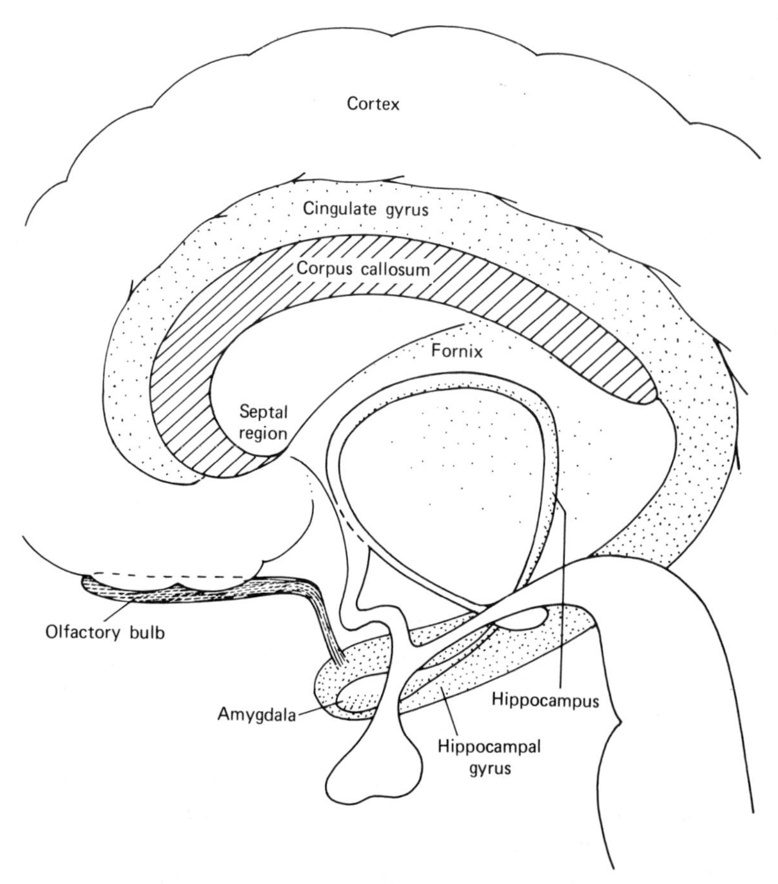

FIGURE 14-4. The principal structures of the limbic system.

have allowed similar experiments to be performed on themselves. By contrast, removal of the septal region renders formerly tame animals ferocious and difficult to manage.

Finally, one of the limbic centers, the hippocampus, also appears to be involved in short-term memory. Patients with lesions in this region appear unable to retain new memories even though already established long-term memories are unimpaired. Apparently in such cases the normal progress of short-term memory into long-term memory is interrupted by hypothalamic lesions. The significance of this finding is not yet well understood.

15
THE BASAL GANGLIA

When the clinician finds a history of tremor of the extremities with normal deep reflexes, increased rigidity of the extensors, a masklike facial expression, and difficulty in initiating walking followed by a rapid, stumbling pace that the patient has trouble stopping, he strongly suspects Parkinson's disease or paralysis agitans (Case XI). This disabling disorder reflects underlying disease of the basal ganglia, a complex collection of subcortical nuclei that, with the cerebral cortex and cerebellum, control motor functions. To carry out their functions, the basal ganglia are richly interconnected with the cerebral cortex, nuclei of the thalamus, and the reticular formation with which they form feedback loops or circuits that go by way of the corticospinal tracts and influence impulses to the muscles. Like the cerebellum and thalamic nuclei, the basal ganglia are not executive centers. They do not initiate movement. Rather, they inhibit excessive movements and lend smoothness to action generated by the motor areas of the cerebral cortex.

ANATOMICALLY SPEAKING: STRIPES, PALE GLOBES, AND STONES

Knowing nothing of the functions of the basal ganglia, early anatomists named them for their appearance, and to have their names before us in order to relate function to structure, we shall begin with that ancient tradition.

The major components of the basal ganglia (Fig. 15-1) are: (1) the caudate (*cauda,* tail) nucleus, an elongated nucleus with a large head region and a long tail that passes backward along the side of the thalamus; (2) the putamen (*putamen,* peach stone), a nucleus that forms part of the lenticular (*lenticularis,* like a lentil) nucleus along with the globus pallidus, which is a cluster of cell bodies transversed by myelinated fibers giving it a pale appearance. Because the area around the caudate nucleus and putamen has a striped or striated appearance, the entire region is often referred to as the corpus striatum, or striated body.

CONNECTIONS OF THE BASAL GANGLIA: LOOP THE LOOPS

The afferent flow to the corpus striatum (Fig. 15-2) arrives from several sources. First, neurons from all parts of the cerebral cortex, but particularly from the motor area, stream into the basal ganglia. Second, the intralaminar nuclei of the thalamus project primarily to the caudate nucleus and putamen. Third, the substantia nigra projects to all nuclei of

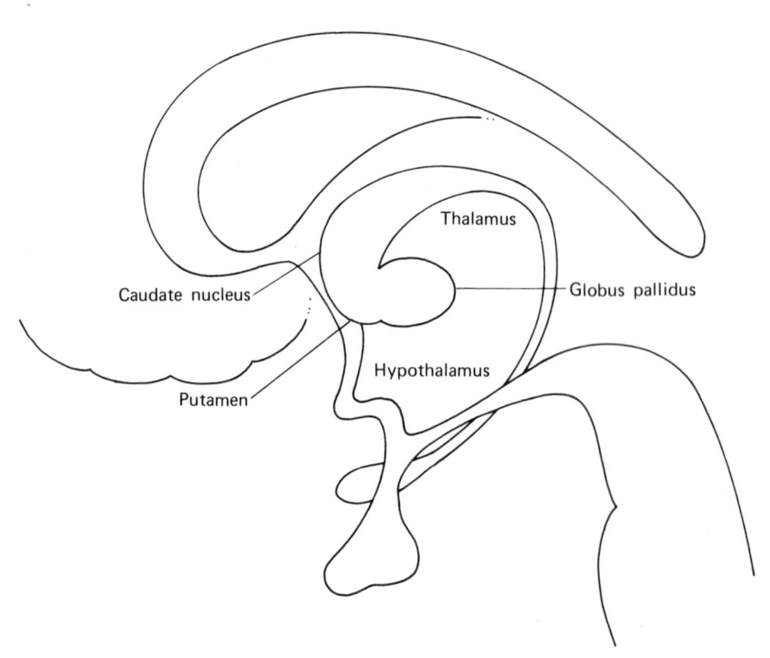

FIGURE 15-1. The principal components of the basal ganglia.

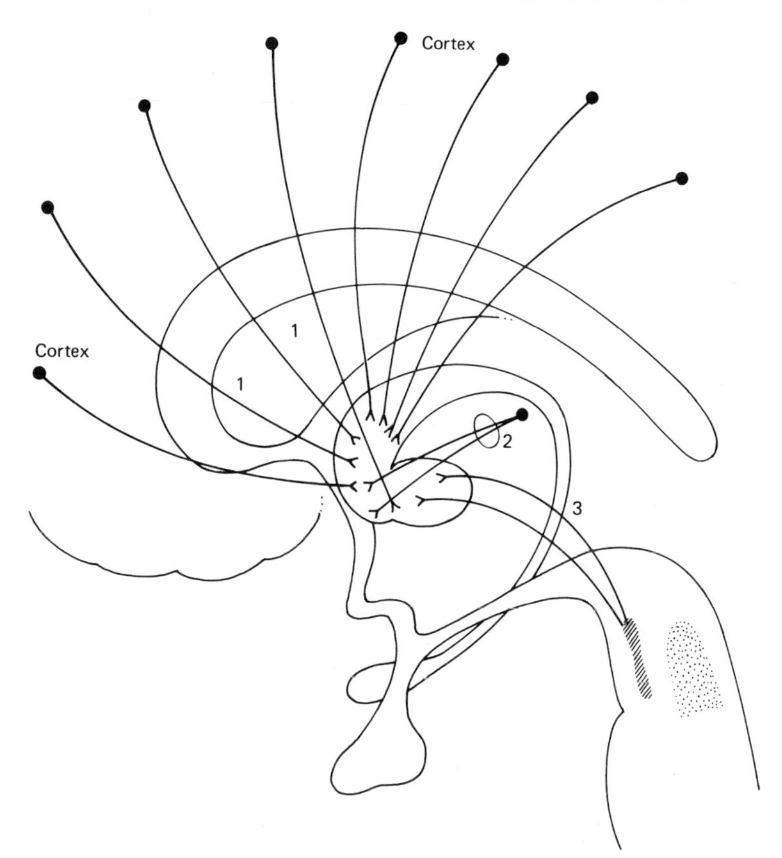

FIGURE 15-2. Afferent input to the corpus striatum. 1, Cortex to corpus striatum; 2, Thalamus to corpus striatum; 3, Substantia nigra to corpus striatum.

the corpus striatum. Significantly the fibers of the substantia nigra transmit dopamine, an inhibitory substance, the significance of which will be discussed later.

The chief efferent outflow from the basal ganglia (Fig. 15-3) arises from the globus pallidus, which projects fibers to the thalamic nuclei, reticular formation, and indirectly to the cerebral cortex through connections to the ventral lateral and ventral anterior nuclei of the thalamus. In addition to the primary efferent outflow to the thalamus, the basal ganglia also have efferent connections to the red nucleus and the rubrospinal tract. These connections allow the basal ganglia to influence the cerebellum and reticular formation.

As both Figures 15-2 and 15-3 show, the influence of the basal ganglia is mediated through loops—circuits that project afferently from

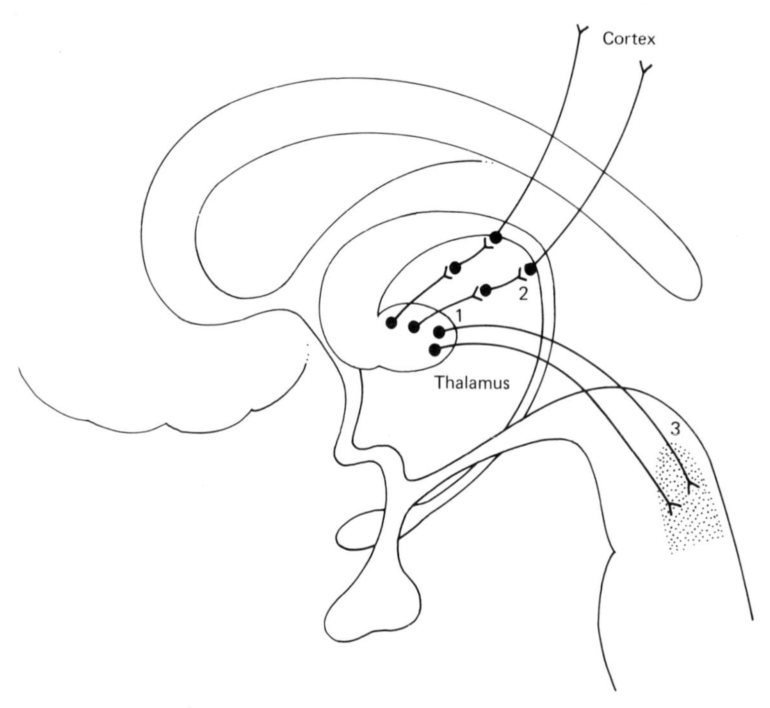

FIGURE 15-3. Efferent output from the basal ganglia. 1, Globus pallidus to thalamus; 2, Thalamus to cortex; 3, Globus pallidus to reticular formation.

the cerebral cortex to the various nuclei of the corpus striatum, which in turn project to each other, to the centers of the midbrain, and then back to the cerebral cortex. Clearly, the ramifications of circuits flowing into and leaving the basal ganglia are such that every pyramidal and extrapyramidal center in the brain comes under their influence in the control of muscular activities.

FUNCTIONS AND MALFUNCTIONS

Based largely on clinical evidence the basal ganglia in man exert their influence on all forms of muscular activity, including posture, balance, and manipulatory and locomotor movements. In general, it appears that their role is one of inhibition of excessive tonus or inappropriate movement. Perhaps the best known and most widely investigated form of pathology of the basal ganglia is Parkinson's disease (Case XII). The chief symptoms of this disorder reflect the inhibitory and selective func-

CASE XII

PARKINSON'S DISEASE

Mrs. B.G., a 72-year-old housewife, developed progressive generalized muscular weakness, rigidity of the extremities, and an unsteady gait. Because of her motor difficulties, she became increasingly dependent on her husband, began to lead a sedentary existence, and suffered from depression. A fall led to her admission to the medical center where diagnostic tests revealed a dull facies (lack of facial expression), general weakness of the musculoskeletal system, rigidity of all extremities, slight slurring of speech, and tremor of the hands at rest, which disappeared with intention to perform movement. She also showed "pill-rolling," or repetitive circular movements of the thumb and index finger against each other.

The patient's gait was shuffling, with difficulty in controlling forward movement. She tended to become propulsive, needing to crash into obstacles in order to stop.

The patient was diagnosed as a case of Parkinson's disease of recent origin and was put on a regimen of Amantadine, a synthetic anticholinergic drug similar in its effects to L-dopa. At the same time she was given a mood-elevating drug because of the depressive side effects associated with the administration of Amantadine.

Within ten days the patient was able to walk without assistance, her tremor and extensor rigidity had disappeared, and she was discharged as capable of maintaining a level of activity normal for her age and general physical condition.

tions of the basal ganglia. The increased tonus, rigidity, jerky movements, tremor, inability to control locomotor movement with the arms failing to swing in rhythm with the legs, all point to a failure of inhibitory mechanisms. Support for this interpretation of the role of the basal ganglia comes from the beneficial effects of L-dopa, which like dopamine, acts as an inhibitory agent on central nervous synapses.

Another disorder associated with the basal ganglia is chorea (*choreia*, a dance or chorus), characterized by a variety of rapid, jerky, and purposeless movements that are, however, well-coordinated. There is no overall change in muscle tonus as is characteristic of Parkinson's disease. One of the better known choreal disorders is Huntington's chorea, an inherited disease in which there is damage to the cerebral cortex as well as to the basal ganglia. Popularly, choreal disorders in children are often referred to as St. Vitus's dance from a medieval patron saint of such diseases.

Atheosis is a disorder characterized by continuous slow, writhing movements that have been described as snakelike or wormlike. To some observers the patient appears to be trying to imitate the sinuous movements of the oriental dance. The movements of atheosis are involuntary and are particularly evident in the hands.

In ballismus the patient engages in wild, violent, flinging movements of the arms—usually of one arm, in which case the disorder is called hemiballismus.

Unfortunately, neither animal experiments nor clinical evidence in man has yielded a consistent picture of the underlying functional mechanisms involved in these disorders. As is so often true in lesions of the brain that result from vascular accidents or tumors, more than one structure may be affected, thus making it difficult to determine the precise basal ganglia involved.

16
THE CEREBRAL CORTEX

The simplest bundle of neurons making up a peripheral nerve and the most complex circuits of the brainstem, basal ganglia, and cerebellum—all are tracts or centers whose fundamental task is the execution of sensorimotor functions. The complex machinery of the body, somatic and visceral, is capable of functioning as an integrated and adaptive whole only through the communications network of the peripheral and visceral nervous system. However, as we began our odyssey through the nervous system (Chap. 1), we cast a hurried glance at the complexities of that highest communications center of all, the cerebral cortex. And it is here that our journey will end. To be sure, the cerebral cortex, like the rest of the nervous system, has its share of sensorimotor functions; for much of our activity is consciously willed by executive centers of the cortex in response to incoming sensory information. However, comparatively little of man's cortex is concerned with basic sensorimotor functions. Rather, the great mass of cortical tissue is responsible for the higher mental processes of learning, memory, thinking, language behavior, music-making, mathematics, and all the rest of those abstract functions so cherished by members of the human species.

Strictly speaking, a cortex is not needed for learning and remembering. Flat worms and octopuses can learn and can prove that they remember what has been learned by repeating the habit. However, species so far down the phylogenetic scale can learn only simple conditioned responses. By contrast, your household's resident dog with his relatively uncomplicated cerebral cortex shows evidence of far more complex learning, thinking, and dreaming than a flat worm or octopus. But again the processes involved remain comparatively simple. Apes with their much more highly developed cortices show reasoning, can use tools and even acquire many concepts from American Sign Language for holding limited conversations with their instructors. Yet even among man's primate cousins there is no spontaneous acquisition of language, no mathematics, science, art, music, strategies of war and peace, and all the other incredibly complex activities of which the human cortex is capable. In this and the chapter to follow, we will focus on this gray mantle that overshadows the rest of the nervous system in complexity of structure and function and that makes so much of our behavior possible.

ARCHITECTURE AND GEOGRAPHY: PYRAMIDS AND STARS

The cerebral cortex lies draped over the cerebrum like a gray, wrinkled mantle (Fig. 16-1). Although it is only a few millimeters thick, the cortex is composed of layers of hundreds of millions of cells intermingled with dendrites and axons. About 90 percent of the cortex is classified as neocortex or isocortex, since it uniformly contains six layers of cells while the remainder contains as few as three layers. Beginning with the surface

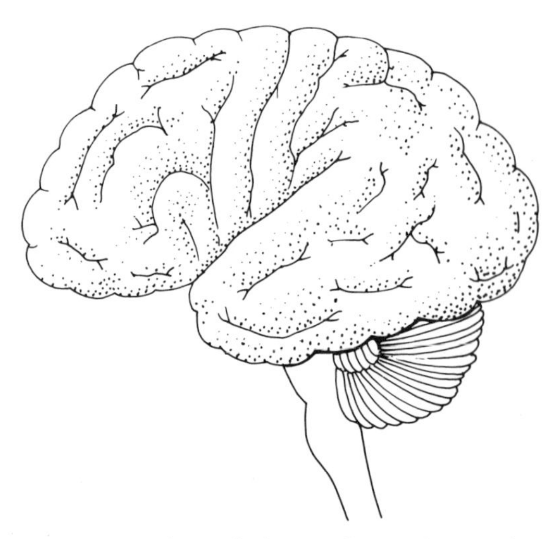

FIGURE 16-1. The cerebral cortex showing the convolutions.

and moving toward the subcortex, the layers of the neocortex have been identified as follows.

I. The molecular, a layer that is not rich in cells but is a thick forest of synaptic fields arising from underlying cells.
II. The external or outer granular, a layer that contains mostly small and medium-sized pyramidal cells. These are shaped like double pyramids and proliferate extensively (Fig. 16-2). This layer also contains Golgi type II cells.
III. The outer pyramidal, a layer that contains large pyramidal cells.
IV. The internal granular, a layer that consists primarily of stellate cells.
V. The internal pyramidal, a layer that contains mostly medium and large pyramidal cells.
VI. The polyform (or multiform), a layer that is made up of a variety of cell types.

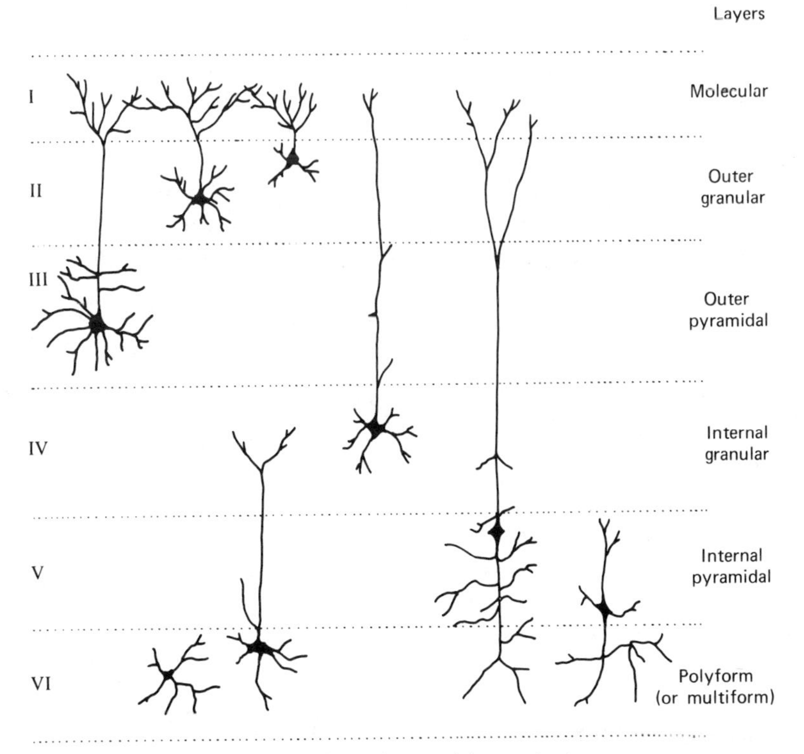

FIGURE 16-2. The six layers of the cerebral cortex.

In general, afferent fibers coming into the cortex synapse in layers I-IV, while the efferent neurons leaving the cortex have their origins in layers V and VI. Of particular importance among the afferent neurons are those from the lateral thalamus that synapse in layer IV. It is also worth noting that the neurons of the cortex are organized not only horizontally in layers but vertically as well. Many of the axons and dendrites of the stellate cells proliferate up and down through all layers of the cortex, their dendritic branches extending for considerable distances horizontally in each layer. And many of the neurons of the cortex are associational in function—that is, they interconnect one area of the cortex with another. Some of these, the commissural fibers, cross from one side of the brain to the other by way of the corpus callosum (*corporis,* body; *callosum,* hard), a large dense band of fibers that connects the right and left cerebral hemispheres. Thus the cortex is clearly organized in such a way as to permit integration and interaction of impulses from all parts of the nervous system both proximal and distal.

In turning from cytoarchitecture to the external geography of the cortex, we are making the transition from microscopic units to crude divisions based on lobes, gyri (*gyrus,* spiral, convoluted), and sulci (*sulcus,* groove, furrow). A gyrus is a strip of tissue lying between two fissures or sulci. The sulci are shallow invaginations of the cortex, while the fissures are deep folds extending for some distance in the cortex.

Figure 16-3 shows the lobes of the brain and identifies the more important sulci and gyri. The numbers on the gyri represent the most important areas of the cortex in a system of classification devised by Karl Brodmann, a German neuroscientist, in 1909. We shall return to these after describing the principal lobes.

On the basis of three principal fissures and sulci, the brain has been divided into four paired lobes. For convenience we shall describe only the lobes on one side of the brain. The frontal lobe includes all the cortex from the central fissure forward. The parietal (*paries,* wall) lobe extends from the central fissure posteriorly to approximately the border of the posterior end of the lateral fissure. Its occipital margins are not clearly marked by a sulcus or fissure. Posterior to the parietal lobe is the poorly defined occipital (*caput,* head) lobe or back of the brain. The temporal lobe lies laterally and ventrally to the lateral fissure. Roughly speaking, the various lobes of the brain have different functions. As we shall see in more detail later, the frontal lobes are concerned with motor functions, the parietal with bodily sensation, the occipital with vision, and the temporal with hearing.

Brodmann's numbering system is an attempt to locate various functional and anatomical regions much more precisely than is possible by

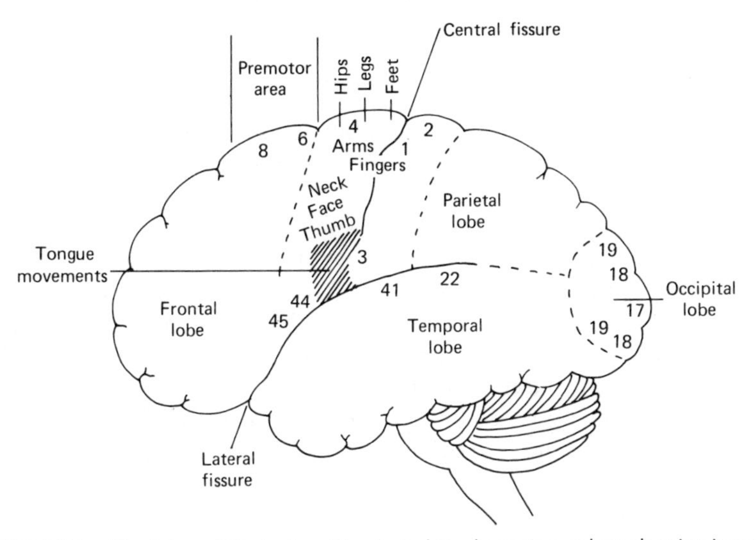

FIGURE 16-3. The lobes of the brain with selected Brodmann's numbers showing important functional areas.

lobes. Each of the major gyri of the cortex has been assigned a number in the system. For example, the gyrus just ahead of the central fissure is labeled 4 in the Brodmann system. This gyrus is the primary motor area. The gyrus just posterior to the central fissure is labelled 3,1,2 and is the somesthetic or bodily sensation area. From time to time we shall specify Brodmann numbers that are in wide use as we describe the various functional areas of the cortex.

MOTOR AND SENSORY AREAS: A STRANGE APPEARING BODY

As we indicated in introducing the Brodmann system, the strip of cortex known as the precentral gyrus (area 4) is the primary motor area. As the term is used here primary means the area is responsible for the initiation of movements. As Figure 16-3 shows, the body musculature is represented in an upside down manner with the dorsal or top part of the area controlling the feet and legs and the bottom part the head. There is, moreover, a disproportionate amount of cortex assigned to the control of the thumb, tongue, and eyes while the trunk has only a small proportion of cortex devoted to its musculature. This arrangement reflects the degree of motility and relative importance in adaptive and manipulative activity of the various parts of the body. The simple bends of which the trunk is capable require little skill compared with the flying fingers of the concert pianist.

As will be remembered from our discussion of spinal pathways in Chapter 3, the motor area of the left hemisphere controls the right side of the body, and the right hemisphere controls the left side of the body. This contralateral control results from the crossing of the corticospinal tracts either in the medulla or in the spinal cord.

Electrical stimulation of the primary motor area generates discrete muscle contractions or relaxations, the particular muscle group affected depending on the place of stimulation. The patient experiences the movement as "willess" or not under his control. Stimulation of the exposed cortex is not painful to the patient whose scalp has been anesthetized before being reflected and a portion of the skull removed. It is often necessary to stimulate the brain during neurosurgery for the control of seizures or the removal of tumors in order to know precisely each region's function. The brain is variable from individual to individual, and for this reason the surgeon must be certain of an area's function before removing it, or the cure may be more debilitating than the disease.

Small lesions of the primary motor area cause flaccid paralysis of contralateral muscle groups with decreased tonus and deep reflexes. However, if the lesion is not too extensive, a remarkable degree of recovery is possible, presumably as a consequence of the utilization of alternate pathways.

The primary area for body sensation, or somesthesis (3,1,2), is located along the postcentral gyrus. Like the motor area it is contralateral in control and upside down in respect to body topography. Electrical stimulation of the area gives rise to false sensations on the opposite side of the body which are referred to the limb or part of the trunk correlated with the cortical region that has been stimulated. Electrical studies also reveal that the somesthetic area and the motor area are closely connected, as might be expected since many forms of response are generated by external stimuli.

Lesions in the somesthetic region cause impairment of touch, pressure, and kinesthesis on the contralateral side. Pain and temperature are affected only slightly. In general, lesions to this region raise the threshold for a sensation rather than abolishing it, with the result that a stronger than normal stimulus is required.

The primary visual area (17) is located on the most caudal portion of the occipital lobe extending in deep between the two hemispheres on the medial surface of the cortex. Because of the crossing of the fibers of the optic tract (Fig. 7-3, p. 125), the right half of each retina is represented on the right visual cortex and the left half on the left cortex. However, because of the refraction of light rays by the lens, the left visual field is represented on the right cortex and the right field on the left cortex. When stimulated in the visual area, the conscious subject experiences points or flashes of light. Destruction of one side of the visual cortex will cause one-half blindness (hemianopsia) in both retinas. If the right cortex is destroyed, the right half of each retina would be blind. Similarly, if the left half were destroyed the left half of each retina would be affected.

The primary sensory area (41) for audition lies along the upper portion of the temporal lobe. In fact, a good proportion of the area is buried within the lateral fissure. Stimulation of the area gives rise to simple tone or noises, while destruction leads to some degree of impairment in both ears, since the auditory fibers from each ear are distributed in both cortical areas.

The primary area for taste has not been identified for certain. It is believed that the second order neurons arising from the solitary nucleus ascend to the ventrobasal thalamic region. If so, then third order neurons would project from there to the somesthetic cortex.

The primary olfactory area lies immediately adjacent to the termination of the olfactory bulbs. The region is called the lateral olfactory area or prepyriform cortex. The pathways and receptive areas are uncrossed and are unique among the sensory pathways in that the olfactory fibers do not project to the thalamus before terminating in the cortex.

THE ASSOCIATION AREAS: FROM SENSING TO PERCEIVING

Each of the primary areas—motor and sensory—has closely associated secondary or association areas (Fig. 16-3). In the case of the motor area, the adjacent secondary areas are those labelled 6 and 8. These areas are often collectively referred to as the premotor area. Area 8 functions in head, neck, and eye movements, particularly those movements involved in tracking a visual stimulus, and for this reason has been called the frontal eye field. Area 6 appears to be concerned with complex motor skills. Because area 6 and area 8 receive rich input from the cerebellum and motor nuclei of the thalamus, these areas may have extrapyramidal functions since the cerebellum and thalamic nuclei are part of this system.

The sensory association areas (Fig. 16-3) are relatively extensive and not precisely delineated. For the somesthetic region a large association area lies posterior to it. Two important visual association areas are 18 and 19, just anterior to the primary visual area (17). Finally, the auditory association area lies around the primary auditory region in the temporal lobe.

The function of the association areas is that of integration, storage of information, and correlation among the various sensory modalities. If, for example, the visual association area is destroyed by a lesion, the individual suffers from visual agnosia (*a,* without; *gnosis,* knowledge) or the inability to recognize objects by vision. Although able to see objects, agnosics cannot identify them by name. Animals trained to make complex visual discriminations will be unable to do so if areas 18 and 19 are destroyed. If the lesion is in the somesthetic area, the individual suffers from the inability to discriminate objects by touch. Such an individual is said to suffer from astereognosis. This condition may be diagnosed by asking the blindfolded individual to pick out common geometric solids from a number of samples. A normal subject can do this easily; the victim of an extensive parietal lesion cannot.

One of the most complex association areas is a relatively large one in the temporal region between the auditory and visual association areas. When stimulated in conscious patients, this region gives rise to complex memories, hallucinations, and even dreams. The famous Canadian neurosurgeon, Wilder Penfield, discovered these functions while exploring the cortices of patients suffering from epileptic seizures. His explorations were directed toward pinpointing the focus of the seizures so that the area could be excised. By stimulating specific points in the temporal cortex with weak electric currents, Penfield could evoke specific and often complex memories in his patients.

THE PREFRONTAL AREA:
A CHANGE IN PERSONALITY

The large mass of tissue ahead of the premotor area is called the prefrontal region. From clinical cases involving lesions in the area, we know that it is involved in planning, management, judgment, and emotional control. Over a century ago an unlucky Vermont railroad worker, Phineas Gage, became the unwilling subject of an accidental destruction of the frontal lobes when a large red-hot metal rod was blown through his head in an explosion at a forge. In effect, he suffered a prefrontal lobotomy. Prior to the accident the subject had been described as a God-fearing man, reliable, moral, and conservative in his habits. But after recovering from his remarkable injuries he underwent a personality change. He was given to profanity, was unreliable in his work habits, used liquor injudiciously, consorted with women of low character, and showed poor judgment in conducting his monetary affairs.

Because accidental injuries to the frontal areas were found to relieve anxiety and to improve the mood of depressed subjects, psychotic patients who showed agitated anxiety with deep depression suggesting over-control of emotionality, have been subjected to prefrontal lobotomy in an attempt to relieve their symptoms. The neurosurgeon inserted a knife under the frontal cortex and swept across the brain, severing the cortex from the underlying structures. After undergoing such an operation, patients showed considerable relief from depression and anxiety, but many underwent undesirable changes in personality, losing initiative, showing poor judgment and lowered moral standards. Partly because of these undesirable side effects and partly as the result of the discovery of tranquilizing drugs, this operation is no longer performed.

THE SPEECH AREA: A ONE-SIDED RELATIONSHIP

In Case I we described the discovery of the speech center by Paul Broca. There we pointed out that speech is typically localized on the left cerebral cortex (areas 44, 45). Broca's discovery was the first substantial piece of evidence that one side of the brain is dominant over the other. Lesions to Broca's area leave the individual with aphasia (*a,* without; *phanai,* to speak). The term is somewhat misleading since the aphasic individual is not literally without the ability to speak; rather he or she speaks only with difficulty, sentences are agrammatical, lacking in many parts of speech ("telegraphese"), and are unintelligible. The ability to express himself or herself in writing may also be impaired.

Some years after the discovery of Broca's area, Karl Wernicke (1848–1905) identified another cortical region (area 22) as intimately concerned with language ability. This region has since been known as Wernicke's area. The area is typically more fully developed in the dominant hemisphere. Lesions to Wernicke's area cause a loss of comprehension of the spoken word. Voices may be heard, but the sounds are without meaning. Moreover, the patient's speech, though fluent, is full of errors, substitutions, and meaningless words and phrases. The individual suffering from this type of disorder seems not to comprehend his own speech and therefore cannot monitor it properly. Such people are also likely to show difficulty in understanding written language.

EXPERIMENT IX

SPLITTING THE BRAIN

A few persons suffering from severe and disabling epileptic seizures that originate in one hemisphere of the brain and spread to the other over the corpus callosum can be treated successfully by an operation in which the corpus callosum is severed. This procedure can result in dramatic improvement in the patient's condition and at the same time leave no serious aftereffects. Such split-brain patients appear to be able to function as well as persons whose hemispheres are connected.

The split-brain subject provides the neuroscientist with a unique opportunity for investigating the role of the right and left hemispheres in language and communication. Roger Sperry and his colleagues at the California Institute of Technology devised a test in which the subject sits before a screen on which a word or picture of a common object can be flashed for a brief period of time (about 1/10 second). The patient meanwhile is required to fixate on a small dot near the center of the screen, which prevents him or her from shifting the gaze from one side of the screen to the other. Using this arrangement, the information from the screen can go only to the opposite hemisphere (see Fig. 7-3 for a diagram of the visual pathways). Thus, if the word "nut" is flashed on the right side of the screen, the information will be transmitted to the left cerebral hemisphere. Similarly, a word or picture shown on the left side of the screen will be transmitted to the right hemisphere.

Under these conditions words or pictures presented in the right visual field can be read and communicated easily. However, when the same stimuli are presented to the left visual field, recognition or communication is difficult or impossible. Particularly interesting is the presentation of words such as "hatband" or "suitcase" by showing the subject one-half of the word in the right visual field and the rest in the left. Only the part of the word that reaches the left hemisphere is recognized and communicated. The subject is forced to guess at the rest of the word.

If the patient is blindfolded or the hand is placed behind a screen, tests of the subject's ability to recognize common objects by touch show that the right hemisphere can recognize objects through this modality, illustrating that the nondominant hemisphere can function in spatial recognition and relations. Interestingly, even though the split-brain subject can recognize an object by touch using the right hemisphere, he or she cannot name it.

Experiments such as these have shown that the main language center is localized in the left hemisphere and nonverbal ideation and spatial relations are localized in the right hemisphere.

From Roger W. Sperry, "Perception in the absence of neocortical commissures," in: *Perception and Its Disorders,* Association for Research in Nervous and Mental Disorders, Vol. 48, 1970.

THE SPLIT BRAIN: ONE HALF DOES NOT KNOW

Experiment IX summarizes what happens when the brain is split. As the report indicates, the main center for language is localized in the left hemisphere. The right hemisphere can comprehend ideas, but these are nonverbal in form. The right or minor hemisphere can also function in stereognosis, or spatial recognition.

We do not know why one side of the brain is dominant over the other. Perhaps linguistic ability, including the highly specialized skill of writing, requires so high a degree of coordination of fine muscle systems that bilateral control would cause conflicts. That neither language facility nor preferential handedness is irreversibly determined for unilateral development is seen in cases of injury to the left cerebral hemisphere in children. They can quickly relearn to speak perfectly and to use the nondominant hand. In adults, retraining is much more difficult, indicating that the development of cerebral dominance is partially a matter of training and habit.

LEARNING: THE CROWNING ACHIEVEMENT

Man's highly developed cerebral cortex makes possible his extensive ability to learn. Of all members of the animal kingdom, human beings are less dependent on instinct than any other species and therefore are more flexible in adapting to their environment.

As we pointed out earlier, the cerebral cortex is not crucial for simple forms of learning such as conditioned responses first studied in Pavlov's famous dogs. However, high level abstract learning does necessitate the presence of a cortex. However, the surprising discovery made and confirmed by several generations of neuroscientists is that the cortex is highly plastic insofar as learning is concerned. That is, learned skills are not precisely localized, and to a surprising degree if one region of the brain is damaged another can function in its place. True, a skill may be temporarily lost, but retraining will usually be possible. The degree of recovery depends upon the extent of the lesion, the level of the subject's ability (very bright people can afford to lose more cortex), and—very importantly—the subject's age. The younger the victim of brain damage, the better the chance for complete recovery.

Finally, we might note that a considerable body of recent research points in the direction that learning is correlated with both chemical and anatomical modification of cortical tissues. Trained animal subjects have both enriched synaptic connections and larger amounts of neuro-transmitters, such as acetylcholine, than animals kept in highly restricted environments. Although this fascinating line of research is still in its infancy, it is promising for our better future understanding of the neurological foundations of learning.

17
DISORDERS OF THE CEREBRUM

WHAT FLESH IS HEIR TO

Throughout the trajectory of life from fetal development to old age, the brain is subject to a variety of abnormalities, diseases, and injuries. Some are the heritage of defective germ plasm, some are congenital or developmental defects that originate during prenatal life, and others are the consequence of diseases, accidents, and aging. Any extended treatment of disorders of the cerebrum would, of course, require a large volume. However, a representative selection of several of the more common types of disorders will serve to illustrate patterns of symptomatology as well as the challenging problems encountered by the neurologist, neurosurgeon, and nursing staff in the diagnosis and treatment of these disorders and the rehabilitation of the patient.

Before considering specific disorders, let us outline certain general problems that are encountered in the evaluation of diseases or injuries to the cerebrum. First, disease processes and traumas are seldom precisely localized in their effects. Tumors, for example, if malignant, progressively invade adjacent tissues or, if benign, create pressure that affect the entire brain. Similarly, penetrating wounds, infections, and vascular accidents usually have consequences beyond the site of primary impact. For these reasons, diagnosis of cerebral disorders may present difficulties of localization, and the clinician typically employs a variety of methods in arriving at a diagnosis. These include observation of behavioral deficits, vital signs, x-rays, EEG records, the injection of dyes into the bloodstream, air into the ventricles, and withdrawal of samples of cerebrospinal fluid.

It is also important to bear in mind that persons with brain injuries may suffer a loss of morale and motivation, realizing as they do the seriousness of neurological disorders. This means that while they may be neurologically capable of performing certain test activities, they may do so only reluctantly or only for special incentives. This consideration becomes of particular importance in nursing and rehabilitative programs.

Finally, it has long been recognized that the same disorder may have different immediate and long-range consequences depending on the individual's age, general physical condition, and levels of intelligence and motivation. While these factors are important in all diseases, they have special impact in neurological disorders, whose consequences tend to be long range.

CONGENITAL HYDROCEPHALUS

A borrowing from the Greek, hydrocephalus means water-headed. It describes the abnormal accumulation of cerebrospinal fluid in the ventricles of the brain, a common congenital defect in infants. Normally the volume of cerebrospinal fluid is in a state of balance, the entire supply being produced and absorbed daily. In congenital hydrocephalus, too much fluid is produced or the pathway for the transmission of the fluid fails to develop or is blocked.

The cerebrospinal fluid is produced by the choroid plexus in the lateral ventricles. From here it passes through the interventricular foramina (Fig. 17-1) to the third ventricle, then through the narrow aqueduct of Sylvius to the fourth ventricle where it leaves the ventricular system by way of the foramina of Luschka and of Magendie to circulate in the spinal cord and to spread along the base of the brain and eventually over the other cerebral surfaces.

Usually in cases of congenital hydrocephalus the head is normal in size at birth, but shortly afterward begins its disproportionate enlargement, often swelling to enormous dimensions. Because the skull of the infant is thin and the sutures not yet closed, there is little resistance to the internal pressure from the accumulating fluid. The specific cause of the accumulation may be a blockage in the cerebral aqueduct or a malformation of the foramina, preventing passage outward from the fourth ventricle. Sometimes the blockage is caused by an elongation of the medulla

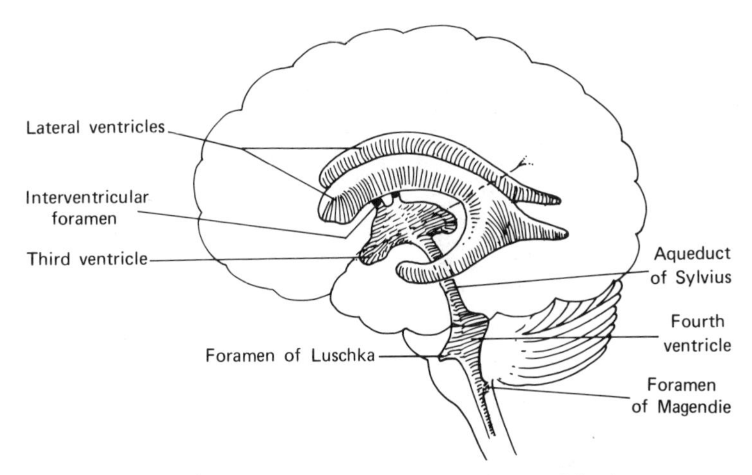

FIGURE 17-1. Pathways for the cerebrospinal fluid.

and cerebellum downward into the foramen magnum preventing the fluid from ascending to the basal portion of the cerebrum and upward. We might note in this connection that hydrocephalus is of two basic types. First, communicating hydrocephalus is the type in which fluid flows readily from the ventricular system into the subarachnoid space but because an excess of fluid is formed, it cannot be absorbed rapidly enough to maintain balance. In noncommnicating hydrocephalus the cause is blockage of the ventricles. In either case, as the fluid accumulates in the ventricles, the spaces enlarge, thinning out the brain tissue. Although sensorimotor functions may remain intact for a relatively long period, mental retardation occurs as a consequence of brain damage. Depending on the severity of destruction, the infant may either die within a few months or survive with mental retardation, spasticity, and ataxia. Some success has been achieved in treating noncommunicating hydrocephalus by establishing an artificial channel from the ventricles into the body cavity. In communicating hydrocephalus, a portion of the choroid plexus may be surgically destroyed, thereby decreasing the amount of fluid being secreted.

HEMATOMA

Hematoma is the name for a swelling filled with extravasated blood. In infants birth injuries to the cranium can cause subdural hematomas with blood accumulating between the cerebral cortex and subdural space (Fig. 17-2). Subdural hematomas also occur in either children or adults

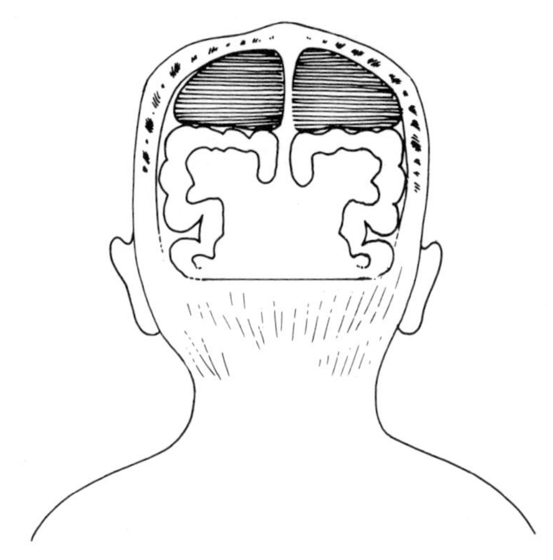

FIGURE 17-2. A bilateral subdural hematoma showing the compression of the cerebrum and distortion of the cranial vault caused by the clot.

from head injuries that damage the delicate arachnoid layer with tearing of small veins. Depending on the size of the hematoma, the symptoms may vary from rapid unconsciousness followed by coma and death, or in less severe cases headache, dizziness, confusion, and unilateral paralysis on the side opposite the lesion and with dilated pupil on the same side as the lesion. Chronic subdural hematomas usually become encysted in a membrane. As the extravasated blood is slowly absorbed and the pressure within the cyst falls, new bleeding or oozing may occur until pressure equilibrium is restored, thus producing variable and fluctuating clinical symptoms. Hematomas are treated by aspiration or suction or by surgical intervention, depending upon the severity of the symptoms. Large hematomas must be removed quickly or compression of the cerebrum will bring on coma and respiratory arrest.

Extradural hematomas occurring between the dura mater and internal surface of the skull are common consequences of skull fractures since the injury to the skull may tear blood vessels in the meninges. A common site of such injuries is in the parietal region where fractures cause shearing of the middle meningeal artery or one of its branches. The arterial hemorrhaging causes rapid compression of the brain with headache, drowsiness, stupor, and finally coma and death. Dilation of the pupil on the same side and paralysis or weakness of the contralateral arm and leg may be present in the early stages of the injury. Treatment consists of opening the skull, removing the hematoma by suction, and tying off the ruptured vessel using special clips or cautery.

Intracerebral hematomas occur because of penetrating wounds or severe head injuries. The symptoms are variable depending on the location and size of the hematoma, but headache, dizziness, vomiting, paralysis, drowsiness, stupor, and coma follow large intracerebral hematomas. Because this type of hematoma is deep within cerebral tissues, treatment is difficult, involving as it does removal of bone fragments, foreign material, and destroyed tissues with control of dural and subdural bleeding.

CEREBROVASCULAR ACCIDENTS

A sudden release of blood into cerebral tissues from a ruptured vessel is known as a cerebrovascular accident (CVA). Lay persons often refer to such accidents as strokes, shocks, or apoplexy. Such accidents are a common cause of sudden death in older individuals or where recovery occurs, of paralysis, contractures, aphasias, and emotional instability.

Cerebrovascular accidents may occur in any part of the brain, and where they involve the region of the pons are rapidly fatal. A common site, less often fatal, involves the middle cerebral artery, which is the terminal branch of the carotid. Radiating over the lateral surfaces of the frontal, parietal, temporal, and anterior portion of the occipital lobes, the middle cerebral artery supplies large masses of cortical tissue with blood. Upon the bursting of one of the larger branches of this artery, the individual collapses, with dilated pupils, respiratory difficulty, paralysis on the contralateral side, and loss of tendon and sphincter reflexes. If coma persists for more than 48 hours, the prognosis is unfavorable, and death usually occurs. If consciousness returns and vital signs remain stable, a slow recovery begins over succeeding months with return of function to paralyzed muscles. However, muscular weakness accompanied by spasticity of movement may persist. Recovery of speech may occur, although some difficulties in articulation persist indefinitely.

One important factor in cerebrovascular accidents is hypertension, the causes of which are not fully understood. Among factors contributing to hypertension are injudicious dietary habits, lack of exercise, and emotional tensions associated with the patient's vocation or home environment. Very often patients may have a history of benign forms of hypertensive accidents with mild transient symptoms, such as headache, dizziness, and sensorimotor disturbances. These are indicative of deficiencies in blood supply in terminal arterioles because of vascular spasms. The surrounding tissues may soften, eventually allowing for the rupture of a major vessel because of decreased pressure in its surroundings. Clearly, a regimen of appropriate treatment is of most benefit in the preaccident stage of hypertension. Changes in diet, reduced intake of fats and salt, the use of tranquilizers, rest, and modified attitudes toward work may reduce hypertension and the likelihood of a major cerebral accident.

CEREBRAL ARTERIOSCLEROSIS

Cerebral arteriosclerosis is typically a progressive degenerative disease of old age. However, the time of onset may be influenced by dietary habits, infectious diseases, or environmental conditions.

The pathological picture of advanced arteriosclerosis of the brain is one of atrophy of gray matter. The cortical sulci are widened giving the convolutions abnormal prominence. Blood vessels are hardened by internal deposits of plaque—hence the common phrase "hardening of the arteries." Irregularly scattered throughout the brain are small areas of softening, a result of spasms of the arterioles.

In the early stages of arteriosclerosis there may be absence of symptoms, and even in advanced cases surprisingly little deficit in sensorimotor or intellectual functions may be observed. However, when the disease becomes severe, the symptomatic pattern is one of impairment of memory (especially for recent memory), dizziness, insomnia, irritability, difficulty in concentration, and emotional instability. The patient becomes careless of his appearance; he or she may be difficult to manage, with disorientation as to time and place. Eventually the victim may be classified as suffering from senile psychosis. There is no cure for arteriosclerosis. The treatment is purely symptomatic, and in the more advanced stages of the disease, involves institutionalization.

GLIOMAS

Tumors of the brain commonly originate in the glia cells and are therefore known as gliomas. Some gliomas are malignant, their neoplastic cells multiplying rapidly and irregularly, invading surrounding neural tissue and blood vessels. Others are nonmalignant, remaining well-demarcated and encysted (Fig. 17-3). However, if nonmalignant tumors create pressure on vital centers, they may be fatal. In the case of a

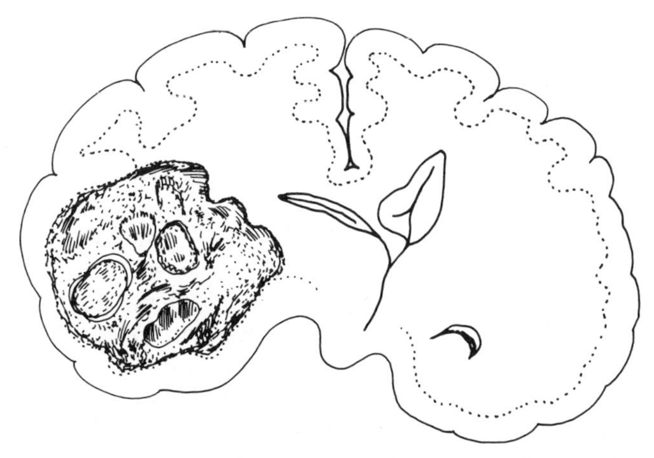

FIGURE 17-3. A glioma of the left cerebral hemisphere showing distortion of the ventricle.

nonmalignant glioma, complete removal of the mass results in a permanent cure. By contrast, malignant types of gliomas spread through the white matter and cerebral cortex causing edema, degeneration of tissue, and hemorrhaging. The patient deteriorates progressively, often rapidly, with convulsions, paralysis, headache, disorientation, and vomiting. In the last stage coma is followed by death. X-ray therapy and surgery may be temporarily beneficial in alleviating symptoms. However, recurrence of the symptoms typically follows from metastisizing cellular masses that escaped surgery.

The two most common types of gliomas are: first, astrocytomas (about 40 percent), which are slow-growing, well-demarcated tumors most frequently occurring in the cerebellar tissues of children or in the cerebrum of the adult. They are characterized histologically by starlike, dark-staining cells. The growing mass causes increased intracranial pressure with varying focalized symptoms depending on the site of the mass. Surgical removal is usually followed by lengthy survival rates. The second type is glioblastoma multiforme, a highly malignant tumor occurring in the cerebral hemispheres. These constitute about 40 percent of the gliomas and usually occur between 40–60 years of age. They originate in the white matter and rapidly advance through the gray matter to the cerebral cortex, where they spread over the surface causing degeneration of neural tissue and scattered hemorrhages. As was indicated earlier, the prognosis is poor and the treatment only palliative and temporarily effective.

INFECTIOUS DISEASES

The brain, like other body organs, is subject to infectious diseases. Among the more common infectious diseases that invade the nervous system are syphilis (see Case IV), rabies, encephalitis, tuberculosis, and meningitis. Because the clinical picture is so complex and variable for infectious diseases, detailed descriptions of the symptomatology and treatment is beyond the scope of this book. In general, it may be observed that the infectious diseases cause progressive inflammation and degenerative changes in brain tissue. In some, such as rabies, the mortality rate is near 100 percent. In others, such as encephalitis, mortality is relatively low, in the order of 10–20 percent. However, upon recovery the patient may show various motor disturbances, and mental and emotional disorders. Where the cause of the disease is bacterial, as in meningitis, syphilis, or tuberculosis, treatment with antibacterial drugs gives good results. When the cause is a virus such as in encephalitis, there may be no known specific treatment. In rabies, where the incubation period is long, serum vaccine containing attenuated viruses induce the production of antibodies (the Pasteur treatment). In nearly all cases, serum vaccine treatment that is early enough prevents mortality, since the disease process is ameliorated or lessened in severity.

GLOSSARY OF ANATOMICAL TERMS

Abducens (*abducere,* to draw outward) The sixth cranial nerve, which turns the eye outward.

Acoustic (*akoustikos,* to hear) The eighth cranial nerve, one of whose branches mediates hearing.

Afferent (*ad,* to; *ferre,* to carry) Pertaining to neurons that conduct impulses toward the central nervous system.

Amacrine fibers (*ama,* vase; *crine,* tentacles) Interconnecting cells in the retina that are shaped somewhat like an octopus.

Amygdala (*amygdala,* almond) An almond-shaped nucleus in the frontolateral region of the cerebrum. It is part of the limbic system.

Aqueous humor (*aqua,* water; *humor,* moisture) The watery fluid that fills the anterior chamber of the eye.

Arachnoid (*arachoneides,* weblike) The middle layer of the three meninges or membranes that cover the central nervous system.

Arbor vitae cerebelli (*arbor,* tree; *vita,* life) Pertaining to the appearance of sagittal sections of the cerebellum.

Archicerebellum (*archein,* oldest, first) The phylogenetically oldest part of the cerebellum.

Arcuate nucleus (*arcus,* bow) A nucleus of the thalamus that serves as a way station for taste impulses on their way to the cerebral cortex.

Axon (*axon,* axis) The elongated and relatively unbranched part of a neuron.

Brachium conjunctivum (*brachion,* arm; *conjunctus,* joined) The superior cerebellar peduncle.

Brachium pontis (*brachion,* arm; *pontis,* bridge) The bridgelike peduncles connecting the two cerebellar hemispheres.

Cauda equina (*cauda,* tail; *equus,* horse) The bushy termination of the spinal cord in the lumbar and sacral regions.

Caudate nucleus (*cauda,* tail) An elongated nucleus in the forepart of the cerebrum that is part of the basal ganglia.

Chiasma (*chiasein,* to mark with an X) A crossing of nerve fibers, such as is found in the optic tracts.

Chorda tympani (*chorda,* string; *tympanos,* drum) A nerve branch made up of fibers from the anterior part of the tongue mediating taste.

Choroid (*choroeides,* chorion or placental tissue) The vascular and darkly pigmented middle layer of the eye.

Cingulate gyrus (*cingulum,* a band or girdle; *gyros,* a circular or spiral form) The part of the cerebral cortex that lies deep in the longitudinal fissure.

Clava (*clava,* club) A small swelling on the dorsal medulla that marks the underlying nucleus of gracilis.

Cochlea (*kochlias,* a snail) The coiled bony structure of the inner ear containing the sense cells for hearing.

Colliculus (*collis,* mound) One of four moundlike swellings making up the superior and inferior colliculi located on the dorsal portion of the midbrain.

Commissure (*commissura,* a junction) A tract of fibers that connects two corresponding regions on the right and left sides of the nervous system.

Cornea (*cornea,* shield) The modification of the sclera in front of the eye. It serves as a lens.

Corpora quadrigemina (*corpus,* body; *quadrigemina,* four twins) A collective name for the paired superior and inferior colliculi.

Corpus callosum (*corpus,* body; *callosum,* hard) A dense band of fibers connecting the right and left cerebral hemispheres.

Corpus striatum (*corpus,* body; *striatum,* striped) A striped area around the basal ganglia.

Crista (*crista,* crest) A ridge of hair cells embedded in a gelatinous mass that are stimulated by movements of the endolymph of the semicircular canals.

Crus cerebri (*crus,* leg; *cerebri,* the cerebral hemispheres) The ventral portion of the cerebral peduncles.

Cuneate tubercle (*cuneus,* wedge-shaped; *tuberculum,* a swelling) A small swelling on the dorsal medulla overlying the site of the nucleus of cuneatus.

Cupula (*cupula,* dome) The gelatinous mass in which the hair cells of the crista are embedded.

Decussation (*decussatus,* in the form of an X) The crossing of nerve tracts from right to left, or vice versa, on their way to or from the centers of the central nervous system.

Dendrite (*dendron,* tree) The highly branched portion of a neuron near the cell body.

Dermatome (*derma,* skin; *tome,* slice or segment) One of the bodily segments innervated by a spinal nerve.

Diencephalon (*dia,* through; *enkephalos,* brain) The midbrain.

Dura mater (*dura,* hard; *mater,* mother) The outermost of the three meninges or membranes that cover the brain and spinal cord.

Ectoderm (*ectos,* outside; *derma,* skin) The outermost layer of the skin.

Efferent (*e,* away from; *ferre,* to carry) Pertaining to neurons that conduct impulses away from the central nervous system.

Emboliform nucleus (*embolos,* stopper) An elongated cerebellar nucleus lying medial to the dentate nucleus.

Endolymph (*endo,* with; *lymph,* water) The fluid that fills the semicircular canals.

Facial (*facies,* the face) The seventh cranial nerve, which serves facial muscles.

Fasciculus cuneatus (*facis,* bundle; *cuneus,* wedge-shaped) With *fasciculus gracilis* (*gracilis,* slender) one of the two tracts making up the dorsal columns of the spinal cord.

Fastigial nucleus (*fastigi,* peak or top) A nucleus of the cerebellum lying on top of the fourth ventricle.

Flocculonodular lobe (*floccus,* a tuft; *nodus,* knot) The ventralmost lobe of the cerebellum.

Fornix (*fornix,* an arch) An arch-shaped nucleus situated above the hypothalamus. Part of the limbic system.

Fovea (*fovea,* pit) The small depression at the rear center of the retina that is the center of clearest daylight vision.

Frontal lobe The portion of the brain lying ahead of the central fissure.

Funiculus (*funiculus,* little bundle) The bundle of fibers in the spinal cord making up either the anterior or posterior white matter.

Ganglion (*ganglion,* a swelling or enlargement) A cluster of nerve cells, usually outside the nervous system.

Geniculate body (*geniculum,* knee) Characterizing a nucleus of bent shape.

Glia See neuroglia.

Globose nucleus (*globose,* globelike) A globe-shaped midline nucleus of the cerebellum.

Glossopharyngeal (*glosso,* tongue; *pharynx,* throat) The ninth cranial nerve serving the tongue and throat.

Gyrus (*gyrus,* spiral, convolution) A strip of cerebral tissue lying between two fissures or sulci.

Hippocampus (*hippo,* horse; *kampos,* sea monster) A portion of the brain next to the temporal horn of the lateral ventricle. It resembles a sea horse. Considered a part of the limbic system.

Hypoglossal (*hypo,* below, *glossa,* tongue) The twelfth cranial nerve situated below the tongue muscles which it innervates.

Hypophysis (*hypophein,* to grow beneath) The pituitary gland located at the base of the brain.

Hypothalamus (*hypo,* below; *thalamus,* chamber) The lower part of the thalamus. Important in the regulation of hunger, thirst, temperature, emotional control, and certain endocrine functions.

Incus (*incus,* anvil) The middle of the three auditory ossicles, named for its resemblance to an anvil.

Infundibulum (*infundere,* to pour) The stalk connecting the pituitary gland to the hypothalamus.

Iris (*iris,* rainbow) The pigmented contractile tissue in front of the eye that controls the amount of light entering the chamber.

Kinesthesis (*kinein,* to move; *esthesis,* feeling) The muscle, tendon, and joint sense.

Lamina (*lamina,* a thin sheet or plate) Pertaining to a vertical sheet of fibers that separates thalamic nuclei.

Lemniscus (*limniskos,* a strip) A band of sensory fibers that ascend the medulla and terminate in the lateral thalamus.

Lenticular nucleus (*lenticularis,* lentillike) One of the nuclei comprising the basal ganglia.

Limbic (*limbus,* border) (1) Pertaining to the limbic lobe, which forms a border around the upper end of the brainstem; (2) pertaining to a complex brain system associated with motivational and emotional processes.

Macula (*macula,* a spot or stain) The structure of the labyrinth of the inner ear containing the receptors for the utricle and saccule.

Malleus (*malleus,* hammer) The auditory ossicle that is connected to the eardrum. It is shaped like an old-fashioned hammer.

Mammillary bodies (*mama,* breast) Two small swellings at the base of the brain that mark the caudal boundary of the hypothalamus.

Medulla oblongata (*medulla,* marrow; *oblongata,* oblong) The lowest part of the brainstem just above the spinal cord. A vital center for respiration.

Meninges (*mennix,* membrane) The membranes covering the brain and spinal cord. They are the dura mater, arachnoid, and pia mater.

Mesencephalon (*mes,* middle; *enkephalos,* brain) The midbrain.

Metencephalon (*met,* after; *enkephalos,* brain) One of the two divisions of the hindbrain. The other is the myelencephalon.

Myelencephalon (*myelos,* marrow; *enkephalos,* brain) One of the two divisions of the hindbrain (see **rhombencephalon**). The other is the metencephalon.

Myelin sheath (*myelos,* marrow) The fatty layer covering larger peripheral neurons.

Myotatic (*mys,* muscle; *tais,* stretching) Characterizing a reflex such as the knee jerk, which is a response to tendon stretch.

Nerve (*nevus,* a sinew) A bundle of neurons.

Neurilemma (*neuron,* nerve; *lemma,* skin) The outer cellular layer surrounding the neuron.

Neuroglia (*neuron,* nerve; *glia,* glue) Supporting cells within the brain. They may also have nutritive functions and act as phagocytes.

Neuron (*nevus,* neuron, a sinew) A single cell of the nervous system

Nucleus (*nuclea,* a nut) (1) The dark staining mass in the center of a cell; (2) a cluster of cell bodies in the central nervous system.

Occipital lobe (*caput,* head) The back or most posterior lobe of the brain.

Oculomotor (*oculus,* eye; *motor,* motion) The third cranial nerve. It innervates muscles of the eye.

Olfactory (*olfacere,* to smell) (1) Pertaining to the olfactory or first cranial nerve serving the sense of smell; (2) pertaining to olfaction or the sense of smell.

Optic (*optikos,* vision) (1) Pertaining to the optic or second cranial nerve, which mediates visual impulses; (2) pertaining to the sense of vision.

Otolith (*ota,* ear; *lith,* stone) Calcium carbonate crystals contained in the macula.

Paleocerebellum (*palaios,* old) The phylogenetically oldest part of the cerebellum.

Parasympathetic (*para,* beyond; *syn,* together; *paschein,* to suffer or feel)

The division of the autonomic nervous system that arises from the cranial and sacral divisions of the central nervous system.

Parietal lobe (*paries,* wall) The lobe of the cerebrum extending from the central fissure to approximately the border of the posterior end of the lateral fissure.

Peduncle (*peduncle,* little foot) A stem in the shape of a foot consisting of fiber tracts connecting one part of the brain to another, such as the pons to the cerebellum.

Perilymph (*peri,* around; *lymph,* water) The fluid that surrounds the semicircular canals.

Phagocytes (*phagos,* eating; *kytos,* cell) Cells capable of ingesting and destroying foreign matter or other cells.

Pia mater (*pia,* tender; *mater,* mother) The innermost of the three meninges or membranes covering the central nervous system.

Pinna (*pinna,* wing) The external ear. The auricle.

Plexus (*plexus,* a turning or folding) A network of fibers.

Pons (*pons,* bridge) The pons consists of a large bundle of fibers connecting the two cerebellar hemispheres, ascending and descending tracts, and nuclei of the cranial nerves.

Prosencephalon (*pro,* forward; *enkephalos,* brain) The first or foremost of the primary embryological vesicles that form the brain.

Putamen (*putamen,* peach stone) A nucleus that forms part of the lenticular nucleus, one of the nuclei of the basal ganglia.

Pyriform lobe (*peri,* pear) A pear-shaped lobe of the cerebrum near the olfactory bulbs.

Ramus (*ramus,* branch) One of the special nerve trunks that attach the spinal nerves to the sympathetic ganglia.

Reflex (*re,* back; *flectere,* to bend) A simple, unlearned response to a stimulus.

Restiform body (*restis,* rope) The inferior cerebellar peduncle.

Reticular formation (*reticulum,* little net) A loosely defined network of fibers extending from the upper spinal cord through the medulla, pons, and brainstem for alerting the cerebral cortex.

Retina (*rete,* net) The innermost layer of the eye consisting of the receptor cells for vision.

Rhizotomy (*rhizos,* root; *otomy,* cut) A surgical procedure in which spinal sensory roots are cut to relieve pain.

Rhodopsin (*rhodo,* rose; *opsin,* appearance) Visual purple or the photopigment found around the rods of the retina.

Rhombencephalon (*rhomb,* a parallelogram; *enkephalos,* brain) The embryological vesicle from which the hindbrain develops.

Saccule (*saccus,* a sac) One of the organs of the sense of equilibrium located in the labyrinth of the inner ear.

Sclera (*sklera,* hard) The leathery outermost covering of the eyeball.

Septum (*sapes,* fence) A partition or division between two masses of neural tissue.

Somesthesis (*soma,* body; *esthesis,* feeling) The body senses including tactile, kinesthetic, and internal sensitivity.

Spinal accessory The eleventh cranial nerve. A helper to the vagus.

Stapes (*stapes,* stirrup) The small auditory ossicle shaped like a stirrup that is attached to the oval window of the cochlea.

Stimulus (*stimulare,* to goad) Any factor or change in energy of sufficient intensity to irritate tissue.

Stria terminalis (*stria,* furrow; *terminus,* end) A fiber tract that extends from the amygdala nucleus to the ventromedial nucleus of the hypothalamus.

Striae medullares (*stria,* furrow; *medulla,* marrow) The striped area marking the upper limit of the medulla.

Substantia nigra (*substantia,* substance; *nigra,* black) A dark-appearing area of the midbrain lying between the cerebral peduncles.

Sulcus (*sulcus,* a groove) A shallow fissure or groove on the surface of the cerebral cortex.

Sympathetic (*Syn,* together; *paschein,* to suffer or feel) With the parasympathetic, one of the divisions of the autonomic nervous system.

Synapse (*syn,* together; *hapsis,* joining) The functional junction between two neurons.

Tectorial membrane (*tectum,* roof) A membrane of the inner ear that rests on the hair cells of the organ of Corti.

Tegmentum (*tegmentum,* covering) The dorsal portion of the cerebral peduncles.

Teledendron (*tele,* distant; *dendron,* tree) One of the terminal branches at the end of an axon.

Telencephalon (*tele,* distant; *enkephalos,* brain) The endbrain from which the cerebral hemispheres develop.

Temporal lobe (*temporalis,* temple) The region of the brain lying on the under surface and side ventral to the fissure of Sylvius.

Thalamus (*thalos,* chamber) A mass of gray matter located near the base of the forebrain. It is a relay center for afferent neurons from all sense organs except the olfactory.

Trigeminal (*trigeminus,* three twins or branches) The fifth cranial nerve whose three branches serve the muscles of mastication and other sensory and motor functions in the head.

Trochlear (*trochlea,* pulley) The fourth cranial nerve serving eye muscles whose tendon is arranged in the form of a pulley.

Tuber cinereum (*tuber,* bump; *cenerous,* ashen) The area of the hypothalamus lying between the optic chiasma and mammillary bodies.

Tympanic membrane (*tympanos,* drum) The ear drum.

Utricle (*uter,* a sac or bag) One of the organs of the sense of equilibrium located in the labyrinth of the inner ear.

Vagus (*vagus,* wandering, straying) The tenth cranial nerve, which wanders down into the thoracic cavity and abdomen innervating the viscera.

Ventricle (*ventriculus,* little belly) A cavity in the brain filled with cerebrospinal fluid.

Vermis (*vermis,* worm) The median region of the cerebellum, which resembles a worm.

Vitreous humor (*vitreus,* glass; *humor,* moisture) The gellike fluid filling the posterior chamber of the eye.

SUGGESTED READINGS

Carlson, N. R. *Physiology of Behavior.* Boston: Allyn and Bacon, Inc., 1977. A thoroughgoing account of the neural correlates of the psychological processes.

Curtis, B. A., S. Jacobson, E. M. Marcus. *An Introduction to the Neurosciences.* Philadelphia, W. B. Saunders, 1972. This advanced introduction to the field contains excellent illustrations and clinical applications.

Gardner, *Fundamentals of Neurology,* 6th ed. For many years used as an introductory text at an intermediate level of difficulty.

Katz, B. *Nerve, Muscle, and Synapse.* New York: McGraw-Hill, 1966. A detailed but clearly written account of neuronal physiology.

Ruch, T. C., H. D. Patton, J, W. Woodbury, and A. L. Towe. *Neurophysiology,* 2nd ed.; Philadelphia: W. B. Saunders, 1965. This is a reprint of the section on neurophysiology from the Howell-Fulton *Textbook of Physiology.* It is an advanced and authoritative account written by experts in various special fields in neurophysiology.

INDEX